SpringerBriefs in Mathematics

W0230295

SpringerBriefs present concise summaries of cutting-edge research and practical applications across a wide spectrum of fields. Featuring compact volumes of 50 to 125 pages, the series covers a range of content from professional to academic. Briefs are characterized by fast, global electronic dissemination, standard publishing contracts, standardized manuscript preparation and formatting guidelines, and expedited production schedules.

Typical topics might include:

- A timely report of state-of-the art techniques
- A bridge between new research results, as published in journal articles, and a contextual literature review
- A snapshot of a hot or emerging topic
- An in-depth case study
- A presentation of core concepts that students must understand in order to make independent contributions

SpringerBriefs in Mathematics showcases expositions in all areas of mathematics and applied mathematics. Manuscripts presenting new results or a single new result in a classical field, new field, or an emerging topic, applications, or bridges between new results and already published works, are encouraged. The series is intended for mathematicians and applied mathematicians. All works are peer-reviewed to meet the highest standards of scientific literature.

Titles from this series are indexed by Scopus, Web of Science, Mathematical Reviews, and zbMATH.

Naohiko Kasuya

Non-Kähler Complex Surfaces and Strongly Pseudoconcave Surfaces

 Springer

Naohiko Kasuya
Department of Mathematics
Hokkaido University
Sapporo, Japan

ISSN 2191-8198   ISSN 2191-8201   (electronic)
SpringerBriefs in Mathematics
ISBN 978-981-96-3001-1   ISBN 978-981-96-3002-8   (eBook)
https://doi.org/10.1007/978-981-96-3002-8

This Springer imprint is published by the registered company Springer Nature Singapore Pte Ltd.
The registered company address is: 152 Beach Road, #21-01/04 Gateway East, Singapore 189721,
Singapore

If disposing of this product, please recycle the paper.

Preface

Compact complex surfaces, Kähler manifolds, and strongly pseudoconvex manifolds are the central objects in complex geometry and the theory of functions of several complex variables. These subjects have been widely studied by many mathematicians and there are a lot of excellent results on them. Our main themes, though, are non-Kähler complex surfaces and strongly pseudoconcave surfaces, which might be regarded as side subjects. However, as the recent results of the author and his collaborators (the construction of non-Kähler complex structures on $\mathbb{R}^4$ [29] and the handlebody construction of strongly pseudoconcave surfaces [88]) show, they touch on many important examples and techniques in 4-dimensional topology, symplectic and contact geometry, as well as complex geometry. These researches have just begun, and are expected to develop in relation to such a wide range of areas of geometry. In this book, we assume researchers in the field of geometry and topology (especially young people, including graduate students) as the main readers and explain our results in order to help them engage with these topics.

Before that, however, it is necessary to first summarize the classical results. The construction in [29] is based on a certain example of a torus fibration related to the theory of genus one Lefschetz fibrations, whose origin is Kodaira's elliptic surface theory. The construction in [88] is also based on several previous works, such as Eliashberg's handlebody construction of Stein manifolds, the convex surface theory by Giroux and Honda (also by Kanda), and the contact surgery theory by Ding–Geiges in 3-dimensional contact geometry. We need to review all these works to some extent. The theory of Lefschetz fibrations belongs to differential topology, and the results by Eliashberg, Giroux, Honda, Kanda, and Ding–Geiges belong to contact geometry. It is more difficult to survey Kodaira's elliptic surface theory, since knowledge of algebraic and complex geometry is needed to understand it. Knowing the basic materials, however, one can grasp a good part of the theory, since the series of his papers [96–102] consists of careful stacking of smart but elementary arguments. Looking ahead the future development of the research, it is important to make it clear that the theory is accessible also from the perspective of a topologist.

After overviewing a broad range of subjects, we will explain how they are used in our results, and show examples of the application of differential topology and contact geometry to complex geometry. This will enable the reader to understand where our results stand between classical complex geometry and modern geometric trends.

Now let us explain the organization of the book.

In Part I (Chaps. 1, 2, 3, and 4), we set elliptic surfaces and non-Kähler complex surfaces at the center of the story, and overview approaches from both complex geometry and differential topology. First in Chap. 1, as a preliminary to Kodaira's elliptic surface theory, we give a brief review of the minimum basics of complex geometry, such as divisors, line bundles, Chern classes, the adjunction formula, and the Riemann–Roch theorem. Next, in Chap. 2, we overview the classification of compact complex surfaces. Especially, we survey the theory of elliptic surfaces and various compact non-Kähler surfaces. More concretely, we deal with the classification by algebraic dimension, the classification of singular fibers in elliptic surfaces, multiple fibers and logarithmic transformations in Sects. 2.2 and in Sect. 2.3 we focus on compact non-Kähler surfaces such as Hopf surfaces, Kodaira surfaces, Inoue surfaces, and Inoue–Hirzebruch surfaces. These are important examples of surfaces of class VII, whose constructions are explicit and attractive from the geometric and topological aspects. In Chap. 3, we see the classification of diffeomorphism types of elliptic surfaces using Lefschetz fibrations. The case where the base space is $\mathbb{C}P^1$ was solved by Kas and Moishezon (Sect. 3.2), and the general case by Matsumoto (Sect. 3.3). This is a good example of the contribution of differential topology to complex geometry. In Sect. 3.4, we introduce the notion of achiral Lefschetz fibrations, a generalization of Lefschetz fibrations. Originally, it was introduced by Matsumoto in order to deal with 4-dimensional topological torus fibration apart from the complex geometry. In fact, however, an example of genus one achiral Lefschetz fibration from S^4 to S^2, called the Matsumoto–Fukaya fibration, plays a very important role in the construction of complex structures in Sect. 4.3. Then, in Chap. 4, as a construction of non-Kähler complex structures on $\mathbb{R}^{2n}$, we explain Calabi–Eckmann's construction ($n > 2$) in Sect. 4.2, and Di Scala–Kasuya–Zuddas' construction ($n = 2$) in Sect. 4.3. The former one is a direct application of the so-called Calabi–Eckmann manifold, which is a holomorphic elliptic bundle from $S^{2p+1} \times S^{2q+1}$ to $\mathbb{C}P^p \times \mathbb{C}P^q$. On the other hand, the approach of the latter one is quite different. By combining the Matsumoto–Fukaya fibration with Kodaira's model of the neighborhood of an I_1 singular fiber, we construct a non-Kähler complex surface diffeomorphic to $\mathbb{R}^4$ together with a holomorphic fibration onto $\mathbb{C}P^1$. Namely, the fusion of the elliptic surface theory and the 4-dimensional topology provides a complex surface that is essentially open, but has an aspect like elliptic surfaces. The obtained non-Kähler complex surface has various remarkable features, which we explain in detail in Sect. 4.4.

Another aim of this book is to show the flexibility of strongly pseudoconcave surfaces. Here it means the flexibility of the topology of complex surfaces and the boundary contact structures. This is in contrast to that of strongly pseudoconvex surfaces, which is very restrictive.

In Part II (Chaps. 5, 6, and 7), setting strongly pseudoconvex (or pseudoconcave) surfaces and their boundary contact structures as the axis, we overview the approaches from both the theory of functions of several complex variables and contact geometry and see what the fusion of them brings. First, in Chap. 5, we review strongly pseudoconvex manifolds and strongly pseudoconvex CR manifolds, so most parts of this section are about known results in the theory of functions of several complex variables. After recalling the definition of pseudoconvexity and the classical results for Stein spaces and Stein manifolds in Sects. 5.1 and 5.2, we overview Eliashberg's handlebody construction of Stein manifolds in Sect. 5.3, which is the only topological and contact geometric part in this chapter. In Sects. 5.4 and 5.5, we deal with strongly pseudoconvex CR manifolds and their fillability, which are central topics in the theory of functions of several complex variables. Next, in Chap. 6, we overview the basics and known results in the field of contact geometry. The main purpose of this chapter is to introduce the contact Dehn surgery in Sect. 6.4, which is necessary for the construction in Chap. 7. In order for that, we give a brief review of the basic facts about contact structures in Sects. 6.1 and 6.2, and survey the convex surface theory by Giroux, Honda, and Kanda in Sect. 6.3. Moreover, in Sects. 6.5 and 6.6, we summarize the definitions and known results of various symplectic fillings and cobordisms. Finally, based on the knowledge prepared in Chaps. 5 and 6, we introduce Kasuya–Zuddas' handlebody construction of strongly pseudoconcave surfaces in Chap. 7. Concretely, we establish the method of holomorphic 2-handle attachment to the strongly pseudoconcave boundary of a complex surface in Sect. 7.2. Our method is based on that of Eliashberg in Sect. 5.3, but there is an important difference concerning the condition on the attaching circle, which makes it possible to realize both (± 1)-contact surgeries by our handle attachments. As a consequence, we can prove that any closed cooriented contact 3-manifold can be realized as the boundary of a strongly pseudoconcave surface. This is in contrast to Bogomolov–de Oliveira's result on the strongly pseudoconvex case and implies the flexibility of the boundary contact structures in the strongly pseudoconcave case.

    The author would like to thank Daniele Zuddas and all other collaborators, since he could not have finished this book without collaborations and discussions with them. He expresses gratitude to Professor Yukio Matsumoto, who told him the background of the construction of the Matsumoto-Fukaya fibration and gave a hint to the concept of the former part of this book. He is also grateful to Professors Kenta Hayano and Yoshihiko Mitsumatsu for their invaluable and helpful comments. Finally, the author wishes to thank Professor Takeo Ohsawa for suggesting that he write this book, and Masayuki Nakamura of Springer for his continuous and professional support. The author acknowledges partial support by JSPS KAKENHI 21K13797.

Sapporo, Japan   Naohiko Kasuya
May 12, 2024

Contents

Part II   Strong Pseudoconvexity, Pseudoconcavity and Contact Structures

Part I
Elliptic Surfaces and Non-Kähler Surfaces

Chapter 1
Preliminaries

In this chapter, we give a brief review of the basics of complex geometry needed for the survey of Kodaira's elliptic surface theory in Sect. 2.2. There, the main important tools will be the long exact sequence of sheaf cohomologies, the Siegel theorem, the Serre duality theorem, the adjunction formula and the Riemann–Roch theorem and its extension. The basic idea of the theory is to describe the quantity of holomorphic sections for various line bundles over a complex curve or surface by sheaf cohomologies, and apply it to the classification of compact complex surfaces. In order to prepare for its usage, we will organize those tools here.

1.1 The Hopf Fibration and the Tautological Line Bundle

To begin with, we explain the tautological line bundle, the first Chern class, and the one point blow-up in relation with the Hopf fibration and the universal S^1-bundle, which are familiar to topologists. Let $(z_0, z_1, \cdots, z_n)$ be the complex coordinates on $\mathbb{C}^{n+1}$, and $[z_0 : z_1 : \cdots : z_n]$ be the homogeneous coordinates on $\mathbb{C}P^n$. Moreover, we regard S^{2n+1} as the unit sphere in $\mathbb{C}^{n+1}$. Let $h_n : S^{2n+1} \to \mathbb{C}P^n$ be the Hopf fibration, that is, h_n is defined by

$$h_n(z_0, z_1, \cdots, z_n) = [z_0 : z_1 : \cdots : z_n].$$

This is canonically a principal S^1-bundle, whose S^1-action is given by

$$e^{i\theta} \cdot (z_0, z_1, \cdots, z_n) = (e^{i\theta} z_0, e^{i\theta} z_1, \cdots, e^{i\theta} z_n).$$

© The Editor(s) (if applicable) and The Author(s), under exclusive license
to Springer Nature Singapore Pte Ltd. 2025
N. Kasuya, *Non-Kähler Complex Surfaces and Strongly Pseudoconcave Surfaces*,
SpringerBriefs in Mathematics, https://doi.org/10.1007/978-981-96-3002-8_1

Then we have the commutative diagram

$$
\begin{array}{ccccccc}
S^3 & \xrightarrow{\ i_1\ } & S^5 & \xrightarrow{\ i_2\ } & S^7 & \xrightarrow{\ i_3\ } & \cdots & \longrightarrow & S^\infty \\
\Big\downarrow{\scriptstyle h_1} & & \Big\downarrow{\scriptstyle h_2} & & \Big\downarrow{\scriptstyle h_3} & & \cdots & & \Big\downarrow{\scriptstyle h_\infty} \\
\mathbb{C}P^1 & \xrightarrow{\ j_1\ } & \mathbb{C}P^2 & \xrightarrow{\ j_2\ } & \mathbb{C}P^3 & \xrightarrow{\ j_3\ } & \cdots & \longrightarrow & \mathbb{C}P^\infty,
\end{array}
$$

where $i_k \colon S^{2k+1} \to S^{2k+3}$ and $j_k \colon \mathbb{C}P^k \to \mathbb{C}P^{k+1}$ are the natural inclusions. Moreover, the *universal S^1-bundle* $h_\infty \colon S^\infty \to \mathbb{C}P^\infty$ is obtained as the inductive limit of h_n.

Next, we see the tautological line bundle. Let $E^{(n)}$ be the $(n+1)$-dimensional complex manifold defined by

$$
E^{(n)} = \{(\lambda x, \pi(x)) \in \mathbb{C}^{n+1} \times \mathbb{C}P^n \mid x \neq \mathbf{0} \in \mathbb{C}^{n+1}, \lambda \in \mathbb{C}\},
$$

where $\pi \colon \mathbb{C}^{n+1} \setminus \{0\} \to \mathbb{C}P^n$ is the canonical projection. Then we obtain the complex line bundle $\gamma_n \colon E^{(n)} \to \mathbb{C}P^n$ given by $\gamma_n(y, z) = z$. This is called the *tautological line bundle* over $\mathbb{C}P^n$. Notice that taking the unit circle in each fiber of this complex line bundle, the Hopf fibration is obtained. In this sense, the corresponding S^1-bundle of γ_n is h_n. Then, just as in the case of the universal S^1-bundle, we obtain the universal complex line bundle $\gamma_\infty \colon E^{(\infty)} \to \mathbb{C}P^\infty$ as the inductive limit of γ_n.

Removing the zero section from the total space $E^{(n)}$ of γ_n, the complement $E_0^{(n)}$ is biholomorphic to $\mathbb{C}^{n+1} - \{0\}$. Therefore, we can consider the operation of removing the origin of $\mathbb{C}^{n+1}$ and inserting the zero section of γ_n, and $E^{(n)}$ can be regarded as the resultant complex manifold. This operation is called a *one point blow-up* or *quadratic transformation*. In fact, this is a local operation. For, even when we restrict both $E_0^{(n)}$ and $\mathbb{C}^{n+1} - \{0\}$ to the parts with radius smaller than r ($r > 0$), we still have a natural biholomorphism. Taking r small, we can do this operation on an arbitrarily small neighborhood of $\mathbf{0} \in \mathbb{C}^{n+1}$. As a consequence, a one point blow-up can be applied at any point in any complex manifold.

Now we explain the first Chern class of a complex line bundle. Using the classifying space $BGL(1; \mathbb{C}) \simeq BU(1) = \mathbb{C}P^\infty$, the first Chern class c_1 is nothing but the canonical generator of $H^2(\mathbb{C}P^\infty; \mathbb{Z}) = \mathbb{Z}$ multiplied by -1. This coincides with the Euler class e of the universal S^1-bundle h_∞, since $\mathbb{C}P^\infty$ is also the classifying space for orientable S^1-bundles, namely, $B\mathrm{Diff}_+(S^1) \simeq BSO(2) = BS^1 = \mathbb{C}P^\infty$. It is well-known that any complex line bundle $\xi \colon F \to X$ can be described as $\xi = f^*\gamma_\infty$ for some smooth map $f \colon X \to \mathbb{C}P^\infty$. Similarly, any orientable S^1-bundle $\eta \colon P \to X$ is the pullback $\eta = g^*h_\infty$ by a smooth map $g \colon X \to \mathbb{C}P^\infty$. Here we define the first Chern class of ξ and the Euler class of η by

$c_1(\xi) = f^* c_1$ and $e(\eta) = g^* e$, respectively. This is the so-called *naturality*. Thus, we have

$$c_1(\gamma_n) = e(h_n) = -1 \in H^2(\mathbb{C}P^n; \mathbb{Z}) = \mathbb{Z}.$$

In other words, denoting by H the generator of $H_{2n-2}(\mathbb{C}P^n; \mathbb{Z}) = \mathbb{Z}$ represented by $\mathbb{C}P^{n-1} \subset \mathbb{C}P^n$, the first Chern class $c_1(\gamma_n) = e(h_n)$ is the Poincaré dual of $-H$ (see Definition 1.1).

Therefore, the exceptional divisor caused by a one point blow-up is an embedded $\mathbb{C}P^n$ with normal Chern class -1, which can be seen as a paraphrase of the fact that the Euler class of the Hopf fibration h_n is -1. When $n = 1$, the normal Chern class of a divisor is identified with its self-intersection number. Hence, the exceptional curve of a one point blow-up is a rational curve (a curve biholomorphic to $\mathbb{C}P^1$) with self-intersection -1. Such a curve is called a (-1) *-curve* or an *exceptional curve of the first kind*. Conversely, if there exists a (-1)-curve on a complex surface, then it can be blown down to one point so that we obtain a new complex surface.

Theorem 1.1 (Castelnuovo's Theorem [19]) *Let X be a smooth complex surface containing a (-1)-curve C. Then there exist another smooth complex surface Y and a holomorphic map $\pi \colon X \to Y$ such that π contracts C to a point p on Y and the restriction $\pi|_{X \setminus C} \colon X \setminus C \to Y \setminus \{p\}$ is a biholomorphism. Moreover, if X is projective, then Y is also projective.*

A compact complex surface is said to be *minimal* if it contains no (-1)-curve. Thanks to Castelnuovo's theorem, we can always obtain from a given compact complex surface a minimal one by a finite iteration of blow-downs.

In 4-dimensional topology, a one point blow-up of a 4-manifold M^4 is usually denoted by $M^4 \# \overline{\mathbb{C}P^2}$. This notation is based on the following observation.

Example 1.1 The complex projective plane $\mathbb{C}P^2$ can be seen as the CW complex obtained by gluing $\mathbb{C}^2$ to $\mathbb{C}P^1$, the divisor at infinity. Now taking a 4-ball $B^4(r)$ of radius r in $\mathbb{C}^2$, then the complement in $\mathbb{C}P^2$ is a tubular neighborhood $N(\mathbb{C}P^1)$ of $\mathbb{C}P^1$. Namely, we have the decomposition

$$\mathbb{C}P^2 = B^4(r) \cup_\partial N(\mathbb{C}P^1).$$

By the tubular neighborhood theorem, $N(\mathbb{C}P^1)$ is diffeomorphic to the total space of the normal complex line bundle of $\mathbb{C}P^1$. Moreover, we have the Hopf fibration h_1 on $\partial B^4(r) = S^3$, while $\partial N(\mathbb{C}P^1)$ admits the same S^1-fibration with the opposite orientation. The two pieces $B^4(r)$ and $N(\mathbb{C}P^1)$ are glued together along these S^1-fibrations by an orientation-reversing diffeomorphism. Therefore, the self-intersection number (=normal Chern class) of $\mathbb{C}P^1 \subset \mathbb{C}P^2$ is $+1$.

On the other hand, if we blow up $\mathbb{C}P^2$ at the origin $\mathbf{0} \in B^4(r)$, the 4-ball $B^4(r)$ is replaced by the total space $E(r)$ of the disk bundle over $\mathbb{C}P^1$ with Euler class (=first Chern class) -1. The resultant closed 4-manifold can be described as

$$E(r) \cup_\partial N(\mathbb{C}P^1).$$

Here the origin $\mathbf{0}$ is replaced by a rational curve with self-intersection number -1, and hence, $E(r)$ is orientation-reversing diffeomorphic to $N(\mathbb{C}P^1)$. Therefore, from the viewpoint of differential topology, a one point blow-up is the same operation as the connected sum with $\overline{\mathbb{C}P^2}$, where $\overline{\mathbb{C}P^2}$ denotes $\mathbb{C}P^2$ with the opposite orientation.

By construction, the 4-manifold $\mathbb{C}P^2 \# \overline{\mathbb{C}P^2}$ obtained in Example 1.1 is naturally a complex surface and admits a $\mathbb{C}P^1$-bundle structure over $\mathbb{C}P^1$. In general, the total space of a $\mathbb{C}P^1$-bundle over $\mathbb{C}P^1$ is called a *Hirzebruch surface* (see Sect. 1.3). In fact, $\mathbb{C}P^2 \# \overline{\mathbb{C}P^2}$ is the only non-minimal Hirzebruch surface. As a smooth 4-manifold, it is diffeomorphic to the total space of the unique non-trivial S^2-bundle over S^2, which we usually denote by $S^2 \tilde{\times} S^2$.

Remark 1.1 There are just two isomorphism classes of smooth S^2-bundles over S^2, since the isomorphism

$$\pi_2(B\mathrm{Diff}_+(S^2)) = \pi_2(BSO(3)) = \pi_1(SO(3)) = \mathbb{Z}_2$$

holds. Here the first equality follows from Smale's result that the diffeomorphism group $\mathrm{Diff}(S^2)$ is homotopy equivalent to $O(3)$.

1.2   Complex Vector Bundles and the Chern Classes

In this section, we give a minimal review of fiber bundles and vector bundles. For more details, the reader is referred to [166] and [76].

Let M, B and F be C^∞-manifolds. A smooth surjective map $\pi : M \to B$ is called a *fiber bundle with fiber F* or an *F-bundle* if there exist an open covering $\{U_\alpha\}$ of B and diffeomorphisms $\varphi_\alpha : \pi^{-1}(U_\alpha) \to U_\alpha \times F$ such that the restriction $\pi|_{\pi^{-1}(U_\alpha)}$ coincides with the composition $\mathrm{pr}_1 \circ \varphi_\alpha$, where pr_1 is the projection to the first factor. Such a diffeomorphism φ_α is called a *local trivialization*. If $U_\alpha \cap U_\beta \neq \emptyset$, we obtain a diffeomorphism

$$\Phi_{\alpha\beta} = \varphi_\alpha \circ (\varphi_\beta|_{\pi^{-1}(U_\alpha \cap U_\beta)})^{-1} : (U_\alpha \cap U_\beta) \times F \to (U_\alpha \cap U_\beta) \times F,$$

which can be written in the form

$$\Phi_{\alpha\beta}(x, y) = \big(x, g_{\alpha\beta}(x)(y)\big).$$

Here the map $g_{\alpha\beta}: U_\alpha \cap U_\beta \to \text{Diff}(F)$ smoothly depends on $x \in U_\alpha \cap U_\beta$, where $\text{Diff}(F)$ denotes the diffeomorphism group of F. Thus we obtain the family $\{g_{\alpha\beta}\}$, which is called *transition functions*. Clearly, $\{g_{\alpha\beta}\}$ satisfies the so-called *cocycle condition*:

$$g_{\alpha\beta}(x) \circ g_{\beta\gamma}(x) \circ g_{\gamma\alpha}(x) = \text{id}_F \quad \text{for any } x \in U_\alpha \cap U_\beta \cap U_\gamma.$$

Conversely, given a family of smooth functions $h_{\alpha\beta}: U_\alpha \cap U_\beta \to \text{Diff}(F)$ satisfying the cocycle condition, then we can construct an F-bundle whose transition functions coincide with $\{h_{\alpha\beta}\}$, and such an F-bundle is unique in the sense of bundle isomorphism. Here two F-bundles $\pi_1: M_1 \to B_1$ and $\pi_2: M_2 \to B_2$ are said to be *isomorphic* if there exist diffeomorphisms $\tilde{f}: M_1 \to M_2$ and $f: B_1 \to B_2$ such that the following diagram commutes:

$$
\begin{array}{ccc}
M_1 & \xrightarrow{\tilde{f}} & M_2 \\
\pi_1 \downarrow & & \downarrow \pi_2 \\
B_1 & \xrightarrow{f} & B_2,
\end{array}
$$

In this case, $\tilde{f}$ is called a *bundle isomorphism covering f*. On the other hand, let $\{g_{\alpha\beta}\}$ and $\{h_{\alpha\beta}\}$ be two families of smooth functions satisfying the cocycle condition. They are said to be *equivalent* if there exists a family of local smooth functions $\lambda_\alpha: U_\alpha \to \text{Diff}(F)$ such that $\lambda_\alpha^{-1} g_{\alpha\beta} \lambda_\beta = h_{\alpha\beta}$. For any two F-bundles over the same base B, the associated open coverings of B are different in general, but by taking a common subdivision, we can discuss whether their transition functions are equivalent or not. Then, it is easy to prove that the two bundles are isomorphic if and only if their transition functions are equivalent. In this sense, a fiber bundle can be identified with its transition functions, which satisfy the cocycle condition.

The range of each transition function is called the *structure group*. Namely, the structure group of a smooth F-bundle is $\text{Diff}(F)$. Now we consider fiber bundles whose fibers are equipped with some structure by reducing the structure group $\text{Diff}(F)$ to a subgroup G. For example, when the fiber is $F = \mathbb{R}^n$, the structure group of a smooth $\mathbb{R}^n$-bundle is $\text{Diff}(\mathbb{R}^n)$, but we can make all the fibers equipped with structure of the n-dimensional vector space by reducing it to $GL(n; \mathbb{R})$. Such an $\mathbb{R}^n$-bundle is called a *real vector bundle of rank n*. Similarly, if the fiber is $F = \mathbb{C}^n$ and $G = GL(n; \mathbb{C})$, such a $\mathbb{C}^n$-bundle is called a *complex vector bundle of rank n*. Notice that the base manifold of a complex vector bundle is not necessarily a complex manifold. Let ξ and η be complex vector bundles of rank m and n, respectively, over the same base manifold B. Then their *Whitney sum $\xi \oplus \eta$* is defined as the complex vector bundle of rank $(m+n)$ whose transition functions are given by $\{f_{\alpha\beta} \oplus g_{\alpha\beta}\}$, where $f_{\alpha\beta}: U_\alpha \cap U_\beta \to GL(m; \mathbb{C})$ and $g_{\alpha\beta}: U_\alpha \cap U_\beta \to GL(n; \mathbb{C})$ are transition functions of ξ and η, respectively, and $(f_{\alpha\beta} \oplus g_{\alpha\beta})(x)$ denotes the direct sum of the two square matrices $f_{\alpha\beta}(x)$ and $g_{\alpha\beta}(x)$ for each $x \in U_\alpha \cap U_\beta$.

Now let B be a complex manifold. A complex vector bundle of rank n over B is called a *holomorphic vector bundle of rank n* if each transition function

$$g_{\alpha\beta} : U_\alpha \cap U_\beta \to GL(n; \mathbb{C})$$

is holomorphic. The group of isomorphism classes of holomorphic line bundles on a complex manifold X is called the *Picard group* of X, denoted by $\mathrm{Pic}(X)$, where the group structure is given by the tensor product of line bundles. Namely, for two line bundles ξ and η with transition functions $\{f_{\alpha\beta}\}$ and $\{g_{\alpha\beta}\}$, respectively, their product is the line bundle $\xi \otimes \eta$, whose transition functions are given by $\{f_{\alpha\beta} g_{\alpha\beta}\}$.

Definition 1.1 (Chern Class) Let X be a smooth manifold. Then, for any complex vector bundle ξ on X and for any non-negative integer j, the cohomology class $c_j(\xi) \in H^{2j}(X; \mathbb{Z})$ satisfying the following axioms is uniquely determined.

1. $c_0(\xi) = 1$.
2. For any smooth map $f : Y \to X$, we have $c_j(f^*\xi) = f^* c_j(\xi)$.
3. Setting $c(\xi) = \sum_{j=0}^{\infty} c_j(\xi)$, then

$$c(\xi \oplus \eta) = c(\xi)c(\eta) \left(= \sum_{k=0}^{\infty} \sum_{j=0}^{k} \left(c_j(\xi) \smile c_{k-j}(\eta) \right) \right)$$

holds for any two complex vector bundles ξ and η over X.
4. For the tautological line bundle γ_n, we have $c_1(\gamma_n) = -1 \in H^2(\mathbb{C}P^n; \mathbb{Z}) = \mathbb{Z}$.

Then $c_j(\xi)$ and $c(\xi) = \sum_{j=0}^{\infty} c_j(\xi)$ are called the *j-th Chern class* and the *total Chern class* of ξ, respectively. If the rank of ξ is n, then we have $c_j(\xi) = 0$ for $j > n$. Therefore, in this case, we call $c_n(\xi)$ the *top Chern class* of ξ. It is well-known that the top Chern class $c_n(\xi)$ can be identified with the Euler class $e(\xi_{\mathbb{R}})$, where $\xi_{\mathbb{R}}$ is the real vector bundle of rank $2n$ associated with ξ.

There are several ways to show the existence of the Chern classes that satisfy the above axioms. The Chern–Weil construction that uses connections and curvatures [22], and the homotopical approach that uses the universal complex vector bundle of rank n over the classifying space $BU(n)$ (cf. [68]) are well-known. Moreover, we can also construct them inductively by the Gysin sequence and the identification $c_n(\xi) = e(\xi_{\mathbb{R}})$ [127]. We end this section by mentioning the following important fact.

Proposition 1.1 *The isomorphism classes of complex line bundles over a smooth manifold M are completely classified by their first Chern classes in $H^2(M; \mathbb{Z})$. Similarly, those of orientable S^1-bundles over M are classified by their Euler classes in $H^2(M; \mathbb{Z})$.*

1.3 Divisors and Holomorphic Line Bundles

In this section, we explain what a divisor is, and how to associate a holomorphic line bundle with it.

Definition 1.2 (Divisors) For a complex manifold X, a linear combination of its irreducible analytic subsets of codimension 1 with integer coefficients is called a *divisor* on X. Two divisors C and D are said to be *linearly equivalent* if there exists a global meromorphic function h whose zero set and poles (h) coincide with the divisor $C - D$.

For example, let us consider the case where X is a compact Riemann surface Σ. A divisor D on Σ is written

$$D = \sum_{i=1}^{k} m_i p_i$$

with a finite number of points $p_1, \dots, p_k \in \Sigma$ and integers $m_1, \dots, m_k$. In this case, the integer $\deg(D) = \sum_{i=1}^{k} m_i$ is called the *degree* of D.

The *divisor group* of a complex manifold X is the additive group of all divisors on X, and is denoted by $\mathrm{Div}(X)$. Linear equivalence classes of divisors on X also form an additive group called the *divisor class group* of X, which is denoted by $\mathrm{Cl}(X)$.

Given a divisor, we can associate a holomorphic line bundle in the following manner. Let D be a divisor on a complex manifold X, and $\{U_\alpha\}$ an atlas of X. On each coordinate neighborhood U_α, there exists a meromorphic function f_α uniquely up to multiplication of non-vanishing holomorphic functions such that $(f_\alpha) = U_\alpha \cap D$ (when $U_\alpha \cap D = \emptyset$, we take $f_\alpha = 1$). Then, the restrictions of f_α and f_β to $U_\alpha \cap U_\beta$ are both meromorphic functions corresponding to the divisor $U_\alpha \cap U_\beta \cap D$. Hence, setting

$$g_{\alpha\beta} = \frac{f_\alpha}{f_\beta}$$

on $U_\alpha \cap U_\beta$, it is a non-vanishing holomorphic function. Since the family $\{g_{\alpha\beta}\}$ satisfies the cocycle condition, we obtain a corresponding line bundle. Moreover, the equivalence class of $\{g_{\alpha\beta}\}$ does not depend on the choice of $\{f_\alpha\}$. Hence, we may denote the line bundle by $[D]$.

Now we show that two line bundles $[C]$ and $[D]$ are isomorphic when the divisors C and D are linearly equivalent. If there exists a global meromorphic function h such that $(h) = C - D$, then we can take a local meromorphic function $h f_\alpha$ on U_α

so that $(hf_\alpha) = U_\alpha \cap C$. Then the line bundle associated with $\{hf_\alpha\}$ is isomorphic to $[C]$. Since

$$\frac{hf_\alpha}{hf_\beta} = \frac{f_\alpha}{f_\beta} = g_{\alpha\beta},$$

it is also isomorphic to $[D]$. Thus we have the isomorphism $[C] \cong [D]$.

Then a natural homomorphism

$$\Phi \colon \mathrm{Cl}(X) \to \mathrm{Pic}(X); \quad \bar{D} \mapsto [D]$$

is obtained, where $\bar{D}$ denotes the equivalence class of a divisor D. On the other hand, for a line bundle L admitting a meromorphic section s, we can associate a divisor D as follows. Let $\{g_{\alpha\beta}\}$ be the transition functions of L. Then a section s is nothing but a collection of local meromorphic functions s_α satisfying $s_\alpha = g_{\alpha\beta}s_\beta$ on $U_\alpha \cap U_\beta$. Since $g_{\alpha\beta}$ is a non-vanishing holomorphic function, we have $(s_\alpha) = (s_\beta)$ on $U_\alpha \cap U_\beta$. Therefore, the collection of local meromorphic functions $\{s_\alpha\}$ define a single divisor D on X. Since

$$\frac{s_\alpha}{s_\beta} = g_{\alpha\beta},$$

the line bundle $[D]$ is isomorphic to L.

Proposition 1.2 Φ *is injective.*

Proof First, we show that Φ is indeed a homomorphism. Let C and D be divisors with transition functions $\{f_{\alpha\beta}\}$ and $\{g_{\alpha\beta}\}$, respectively. Then, the transition functions of the divisor $C + D$ are $\{f_{\alpha\beta} \cdot g_{\alpha\beta}\}$, which coincide with those of the line bundle $[C] \otimes [D]$. Therefore,

$$\Phi(C + D) = \Phi(C) \cdot \Phi(D).$$

Now let us show the injectivity of Φ. Suppose $[D]$ is the trivial line bundle. Then there exists a non-vanishing holomorphic section on $[D]$, namely, a collection of local non-vanishing holomorphic functions $\{h_\alpha\}$ satisfying $h_\alpha = g_{\alpha\beta}h_\beta$. The collection of local meromorphic functions $\{f_\alpha\}$ associated with the divisor D also satisfies $f_\alpha = g_{\alpha\beta}f_\beta$. Hence, $\{f_\alpha h_\alpha^{-1}\}$ defines a global meromorphic function on X. Indeed, the equality

$$f_\alpha h_\alpha^{-1} = g_{\alpha\beta} f_\beta (g_{\alpha\beta}h_\beta)^{-1} = f_\beta h_\beta^{-1}$$

holds on $U_\alpha \cap U_\beta$. Therefore, the divisor D is linearly equivalent to the zero divisor. $\qquad\square$

It is known that Φ is an isomorphism if X is projective. In particular, any line bundle L over a compact Riemann surface Σ can be written as $L = [D]$ for

some divisor D (see Remark 1.4). In this case, we can describe the divisor as
$D = \sum_{i=1}^{k} n_i p_i$ with $p_i \in \Sigma$ and $n_i \in \mathbb{Z}$. Then the first Chern class of the line
bundle $L = [D]$ is given by

$$c_1([D]) = \deg(D) = \sum_{i=1}^{k} n_i \in H^2(\Sigma; \mathbb{Z}) = \mathbb{Z}.$$

This is obvious from the fact that $c_1(L)$ ($= e(L_{\mathbb{R}})$) is the obstruction to the
existence of a non-vanishing section of L, which we regard here as a complex line
bundle.

Now we introduce an important example of a line bundle called the *canonical
line bundle*. Let X be a complex n-manifold and $\{U_\alpha\}$ its atlas. We take coordinate
neighborhoods U_α and U_β such that $U_\alpha \cap U_\beta \neq \emptyset$, and local coordinates $(z_1, \ldots, z_n)$
and $(w_1, \ldots, w_n)$. Then we obtain non-vanishing holomorphic n-forms $dz_1 \wedge \cdots \wedge
dz_n$ on U_α and $dw_1 \wedge \cdots \wedge dw_n$ on U_β with the relation

$$dw_1 \wedge \cdots \wedge dw_n = \det\left(\frac{\partial(w_1, \ldots, w_n)}{\partial(z_1, \ldots, z_n)}\right) dz_1 \wedge \cdots \wedge dz_n$$

on $U_\alpha \cap U_\beta$, where $\left(\frac{\partial(w_1,\ldots,w_n)}{\partial(z_1,\ldots,z_n)}\right)$ is the Jacobian of the coordinate transformation.
Then we can define a holomorphic line bundle whose transition functions are given
by

$$g_{\alpha\beta} = \det\left(\frac{\partial(w_1, \ldots, w_n)}{\partial(z_1, \ldots, z_n)}\right).$$

This is called the *canonical line bundle* of X, and usually denoted by K_X. We
can also define K_X as the determinant bundle $\det(T^*X) = \bigwedge^n T^*X$, where
T^*X denotes the cotangent bundle of X (therefore, $K_X = T^*X$ if $n = 1$). By
construction, a holomorphic section of K_X is nothing but a global holomorphic n-
form on X.

Now let us compute the degree of K_Σ for a compact Riemann surface Σ of genus
g. Recalling the isomorphism of line bundles $K_\Sigma = T^*\Sigma \cong (T\Sigma)^{-1}$, we obtain the
following:

$$\deg(K_\Sigma) = -\deg(T\Sigma) = -c_1(T\Sigma) = -e(T\Sigma_g) = -\chi(\Sigma_g) = 2g - 2.$$

Here Σ_g denotes a closed oriented surface of genus g, which we obtain from Σ
by forgetting the complex structure. The next simple example will be useful for
reviewing the contents of this section.

Example 1.2 We regard $\mathbb{C}P^1$ as the gluing of two copies of the complex plane $\mathbb{C}$. Namely, setting $p_0 = [1 : 0]$ and $p_\infty = [0 : 1]$, we have an atlas of $\mathbb{C}P^1$ consisting of a coordinate neighborhood $\mathbb{C}P^1 - \{p_\infty\} = \mathbb{C}_z$ containing $p_0 \in \mathbb{C}P^1$ and a coordinate neighborhood $\mathbb{C}P^1 - \{p_0\} = \mathbb{C}_w$ of $p_\infty \in \mathbb{C}P^1$, with the coordinate transformation $w = \frac{1}{z}$. Now we describe the tangent bundle $T\mathbb{C}P^1$ and the cotangent bundle $T^*\mathbb{C}P^1$ as holomorphic line bundles over $\mathbb{C}P^1$. First we take a non-vanishing holomorphic vector field $\frac{\partial}{\partial z}$ on $\mathbb{C}_z$. By the coordinate transformation $w = \frac{1}{z}$, it is written as

$$\left(\frac{dw}{dz}\right)\left(\frac{\partial}{\partial w}\right) = -w^2\left(\frac{\partial}{\partial w}\right)$$

on $\mathbb{C}_w - \{0\}$, and it extends over $\mathbb{C}_w$ as a vector field having the zero of order 2 at 0. Thus we obtain a holomorphic vector field on $\mathbb{C}P^1$ having the zero of order 2 at p_∞. Since we can regard it as a holomorphic section of $T\mathbb{C}P^1$, we obtain the description $T\mathbb{C}P^1 = [2p_\infty]$. As is easily verified, any two points on $\mathbb{C}P^1$ are linearly equivalent to each other. Therefore, we can replace p_∞ by any point $p \in \mathbb{C}P^1$, and write $T\mathbb{C}P^1 = [2p]$. We may even write $T\mathbb{C}P^1 = O(2)$ by omitting p. Hence, it follows that $c_1(T\mathbb{C}P^1) = \deg([2p]) = 2$, which coincides with $e(TS^2) = \chi(S^2) = 2$.

In a similar way, we can compute the degree of $T^*\mathbb{C}P^1$. First we take the non-vanishing holomorphic 1-form dz on $\mathbb{C}_z$. Pulling it back on $\mathbb{C}_w - \{0\}$ by the coordinate transformation $z = \frac{1}{w}$, we obtain the holomorphic 1-form

$$d\left(\frac{1}{w}\right) = -\frac{1}{w^2}dw,$$

which extends over $0 \in \mathbb{C}_w$ meromorphically. Thus we obtain a meromorphic 1-form on $\mathbb{C}P^1$, that is, a meromorphic section of $T^*\mathbb{C}P^1$, having a pole of order 2 at p_∞. Hence, it follows that

$$T^*\mathbb{C}P^1 = [-2p] = O(-2).$$

Therefore, $c_1(T^*\mathbb{C}P^1) = \deg(T^*\mathbb{C}P^1) = -2$ and $c_1(K_{\mathbb{C}P^1}) = \deg(K_{\mathbb{C}P^1}) = -2$.

Finally, we define Hirzebruch surfaces. As in the above example, we take the standard atlas of $\mathbb{C}P^1$ consisting of $\mathbb{C}_z$ and $\mathbb{C}_w$. For any integer n, we consider the holomorphic line bundle L_n over $\mathbb{C}P^1$ with transition function $w^n(= z^{-n})$ on $\mathbb{C}_z \cap \mathbb{C}_w \cong \mathbb{C}^*$. Then it is just $O(n)$. Similarly, $L_{-n} \cong O(-n)$ is given by the transition function $w^{-n} = z^n$. Now, just as in Example 1.1, we can holomorphically glue the total spaces of these two line bundles so that at each point on the base $\mathbb{C}P^1$, the two fibers are glued to form $\mathbb{C}P^1$. Thus, we obtain a holomorphic $\mathbb{C}P^1$-bundle over $\mathbb{C}P^1$. The total space of this bundle is called a *Hirzebruch surface* and denoted by H_n. Notice that $H_0 \cong \mathbb{C}P^1 \times \mathbb{C}P^1$ and H_n is isomorphic to H_{-n} as a holomorphic $\mathbb{C}P^1$-bundle by construction. Hirzebruch, who first introduced these surfaces, proved the following result on them.

Proposition 1.3 (Hirzebruch [70]) *Let m, n be non-negative integers. Then, H_m and H_n are not biholomorphic to each other if $n \neq m$. On the other hand, they are diffeomorphic as smooth 4-manifolds if and only if $n \equiv m$ modulo 2.*

Therefore, H_n is diffeomorphic to $S^2 \times S^2$ if n is even, and to $S^2 \tilde{\times} S^2$ if n is odd.

1.4 Sheaf Cohomology and Exact Sequence

In this section, we review the basics of sheaf cohomology of complex manifolds. A sheaf $\mathcal{F}$ on a topological space X is a correspondence from each open set U of X to an additive group $\mathcal{F}(U)$ which is compatible with the restriction map and satisfies the locality axiom and the gluing axiom. For example, when X is a complex manifold, we have a sheaf O_X that associates each open set U with the ring $O_X(U)$ of all the holomorphic functions over U. This is called the *sheaf of germs of holomorphic functions* on X (for details about sheaves, see [64]).

Let $\mathcal{F}$ be a sheaf on a topological space X and $\mathcal{U} = \{U_\alpha\}$ a locally finite cover of X. Then, the sheaf cohomology group $H^*(\mathcal{U}, \mathcal{F})$ is defined as follows. A k-simplex of $\mathcal{U}$ is an ordered pair of $(k+1)$ open sets $U_{\alpha_0}, \ldots, U_{\alpha_k}$ satisfying $\bigcap_{i=0}^{k} U_{\alpha_i} \neq \emptyset$. If there is a section

$$s_{\alpha_0 \ldots \alpha_k} : \bigcap_{i=0}^{k} U_{\alpha_i} \to \mathcal{F}$$

for each k-simplex $U_{\alpha_0}, \ldots, U_{\alpha_k}$ such that $s_{\alpha_0 \ldots \alpha_k}$ is alternating with respect to $\alpha_0, \ldots, \alpha_k$, the collection of such sections $s = (s_{\alpha_0 \ldots \alpha_k})$ is called a k-*cochain with coefficients in $\mathcal{F}$*. We denote the set of k-cochains with coefficients in $\mathcal{F}$ by $C^k(\mathcal{U}, \mathcal{F})$. The coboundary operator $\delta^k : C^k(\mathcal{U}, \mathcal{F}) \to C^{k+1}(\mathcal{U}, \mathcal{F})$ is defined by

$$(\delta^k c)_{\alpha_0 \ldots \alpha_{k+1}} = c_{\alpha_1 \ldots \alpha_{k+1}} - c_{\alpha_0 \alpha_2 \ldots \alpha_{k+1}} + \cdots + (-1)^{k+1} c_{\alpha_0 \ldots \alpha_k}.$$

Then, $\delta^k \circ \delta^{k-1} = 0$ holds, and hence, the k-dimensional cohomology group in $\mathcal{F}$-coefficients is defined by

$$H^k(\mathcal{U}, \mathcal{F}) = \ker(\delta^k) / \mathrm{Im}(\delta^{k-1}).$$

Moreover, taking the inductive limit about subdivisions of the open cover, we obtain the k-dimensional cohomology group

$$H^k(X, \mathcal{F}) = \lim_{\mathcal{U}} H^k(\mathcal{U}, \mathcal{F}),$$

which does not depend on the open cover $\mathcal{U}$. The cohomology obtained in such a way is called the Čech *cohomology*.

In the following, let us consider mainly the case where X is a complex manifold, and the sheaf $\mathcal{F}$ is either O_X or O_X^*. Here O_X^* denotes the sheaf of germs of non-vanishing holomorphic functions on X. Moreover, we will also consider the sheaf $O(E)$ of germs of holomorphic sections of a holomorphic line bundle E over X. In this case, we will sometimes denote the cohomology group by $H^k(X, E)$ instead of $H^k(X, O(E))$.

First we will reveal the meaning of the 0-dimensional cohomology group $H^0(X, O_X)$. A 0-cochain $s \in C^0(\mathcal{U}, O)$ is, by definition, a collection of local holomorphic functions f_α defined over U_α. It represents an element of $H^0(X, O_X)$ if and only if it is a 0-cocycle, namely, $\delta s = 0$ holds. Recalling that

$$(\delta s)_{\alpha\beta} = f_\alpha - f_\beta,$$

the condition is equivalent to the one that local holomorphic functions f_α and f_β coincide over $U_\alpha \cap U_\beta$. Therefore, we obtain a global holomorphic function f over X from the collection of local holomorphic functions $\{f_\alpha\}$. Thus we have an identification between an element of $H^0(X, O_X)$ and a global holomorphic function over X. In general, $H^0(X, O(E))$ is identified with the space $\Gamma(E)$ of global holomorphic sections of the line bundle E. In particular, $H^0(X, O(K_X))$ is the vector space $\Omega^n(X)$ consisting of global holomorphic n-forms over X, where K_X is the canonical line bundle on X.

Next, we will see the meaning of $H^1(X, O_X^*)$. In fact, an element of this cohomology group is identified with an element of $\mathrm{Pic}(X)$, namely, an isomorphism class of holomorphic line bundles on X. A 1-cochain $c \in C^1(\mathcal{U}, O_X^*)$ is a collection of local non-vanishing holomorphic functions $g_{\alpha\beta}$ satisfying $g_{\beta\alpha} = g_{\alpha\beta}^{-1}$, each of which is defined on $U_\alpha \cap U_\beta$ with $U_\alpha \cap U_\beta \neq \emptyset$. The condition that $c = (g_{\alpha\beta})$ is a 1-cocycle is equivalent to $\delta c = 0$, namely, the equality

$$(\delta c)_{\alpha\beta\gamma} = c_{\beta\gamma} - c_{\alpha\gamma} + c_{\alpha\beta} = 0$$

holds for any α, β, γ such that $U_\alpha \cap U_\beta \cap U_\gamma \neq \emptyset$. Notice here that O^* is regarded as an additive group for multiplication, and hence, the above equality means the following:

$$g_{\beta\gamma} g_{\alpha\gamma}^{-1} g_{\alpha\beta} = 1 \iff g_{\alpha\beta} g_{\beta\gamma} g_{\gamma\alpha} = 1.$$

This is exactly the cocycle condition, which we have seen in Sect. 2.1. Therefore, $\{g_{\alpha\beta}\}$ can be seen as the transition functions of a holomorphic line bundle. Moreover, an element of $\delta C^0(\mathcal{U}, O^*)$ is described as δs with a 0-cochain $s = (f_\alpha)$. Since we have

$$(\delta s)_{\alpha\beta} = s_\beta - s_\alpha \ \left(= \log(f_\beta f_\alpha^{-1}) \right),$$

two transition functions $\{g_{\alpha\beta}\}$ and $\{h_{\alpha\beta}\}$ are equivalent $\left(h_{\alpha\beta}g_{\alpha\beta}^{-1} = f_\beta f_\alpha^{-1}\right.$ for some $s = (f_\alpha))$ if and only if they are cohomologue as 1-cochains. Thus a 1-dimensional cohomology class corresponds to an equivalence class of transition functions, and we have obtained the following group isomorphism:

$$H^1(X, O_X^*) \cong \mathrm{Pic}(X).$$

In the theory of sheaf cohomology, the following is a fundamental fact.

Theorem 1.2 *If there exists a short exact sequence of sheaves on* X

$$0 \to \mathcal{F} \to \mathcal{G} \to \mathcal{H} \to 0,$$

the long exact sequence of cohomologies

$$\begin{aligned}
0 &\longrightarrow H^0(X, \mathcal{F}) \longrightarrow H^0(X, \mathcal{G}) \longrightarrow H^0(X, \mathcal{H}) \\
&\longrightarrow H^1(X, \mathcal{F}) \longrightarrow H^1(X, \mathcal{G}) \longrightarrow H^1(X, \mathcal{H}) \\
&\longrightarrow H^2(X, \mathcal{F}) \longrightarrow H^2(X, \mathcal{G}) \longrightarrow H^2(X, \mathcal{H}) \\
&\longrightarrow \quad \cdots
\end{aligned}$$

is induced.

Here we need to explain what a *sheaf homomorphism* and a *sheaf exact sequence* are. For a sheaf $\mathcal{F}$ on X and a point $x \in X$, the inductive limit

$$\mathcal{F}_x = \lim_{U \to x} \mathcal{F}(U)$$

is called the *stalk* of $\mathcal{F}$ at x. A *homomorphism* between two sheaves $\mathcal{F}, \mathcal{G}$ on X is a collection of group homomorphisms $\mathcal{F}(U) \to \mathcal{G}(U)$ that is compatible with the restriction maps. It induces a homomorphism between stalks $\mathcal{F}_x \to \mathcal{G}_x$ for each $x \in X$. A *short exact sequence of sheaves* is a sequence of sheaf homomorphisms

$$0 \to \mathcal{F} \to \mathcal{G} \to \mathcal{H} \to 0$$

such that for each $x \in X$,

$$0 \to \mathcal{F}_x \to \mathcal{G}_x \to \mathcal{H}_x \to 0$$

is an exact sequences between stalks.

Example 1.3 We show two important examples of short exact sequences of sheaves.

1. For any line bundle L and a divisor D on a complex manifold X, there is a short exact sequence

$$0 \to O(L \otimes [D]^{-1}) \to O(L) \to O(L|_D) \to 0.$$

The exactness is obvious from the fact that a holomorphic section of $L \otimes [D]^{-1}$ is identified with that of L vanishing on D.

2. For any complex manifold X, we have a short exact sequence

$$0 \longrightarrow \mathbb{Z} \stackrel{j}{\longrightarrow} O_X \stackrel{e}{\longrightarrow} O_X^* \longrightarrow 0,$$

where j is the natural inclusion and e is defined by $e(f) = \exp(2\pi i f)$ for any $f \in O_X(U)$. This is called the *exponential sheaf sequence*.

From the exponential sheaf sequence, the long exact sequence

$$
\begin{aligned}
0 &\longrightarrow H^0(X, \mathbb{Z}) \stackrel{j^*}{\longrightarrow} H^0(X, O) \stackrel{e^*}{\longrightarrow} H^0(X, O^*) \\
&\stackrel{\delta}{\longrightarrow} H^1(X, \mathbb{Z}) \stackrel{j^*}{\longrightarrow} H^1(X, O) \stackrel{e^*}{\longrightarrow} H^1(X, O^*) \\
&\stackrel{\delta}{\longrightarrow} H^2(X, \mathbb{Z}) \stackrel{j^*}{\longrightarrow} H^2(X, O) \stackrel{e^*}{\longrightarrow} H^2(X, O^*) \\
&\stackrel{\delta}{\longrightarrow} \quad \cdots
\end{aligned}
$$

is obtained. As is well-known, the homomorphism $\delta : H^1(X, O^*) \to H^2(X, \mathbb{Z})$ in the long sequence is the map that corresponds a line bundle $L \in \mathrm{Pic}(X)$ to $-c_1(L)$. By the exactness of the above sequence, the cokernel of the homomorphism $j^* : H^1(X, \mathbb{Z}) \to H^1(X, O)$ is isomorphic to

$$\ker(\delta) = \mathrm{Pic}^0(X) = \{L \in \mathrm{Pic}(X) \mid c_1(L) = 0\}.$$

Now let X be a compact Riemann surface Σ of genus g. Then, composed with the Serre duality $H^1(\Sigma, O) \cong H^0(\Sigma, \Omega) = \Omega^1(\Sigma)$, the cokernel of the homomorphism $j^* : \mathbb{Z}^{2g} = H^1(\Sigma, \mathbb{Z}) \to H^1(\Sigma, O) = \mathbb{C}^g$ is identified with the Jacobi variety

$$\mathrm{Jac}(\Sigma) = \mathbb{C}^g / \mathbb{Z}^{2g} \cong \mathrm{Pic}^0(\Sigma).$$

Now we have the following short exact sequence:

$$0 \longrightarrow \mathrm{Jac}(\Sigma) \longrightarrow \mathrm{Pic}(\Sigma) \stackrel{c_1}{\longrightarrow} \mathbb{Z} \longrightarrow 0.$$

Hence, by putting $g = 0$, we obtain $\mathrm{Pic}(\mathbb{C}P^1) = \mathbb{Z}$. In other words, holomorphic line bundles over $\mathbb{C}P^1$ are completely classified by their first Chern classes. This coincides with the classification of complex line bundles over $\mathbb{C}P^1$ (see

Proposition 1.1). In contrast, when $g \geq 1$, there are infinitely many pairwise non-isomorphic line bundles with trivial Chern class.

Remark 1.2 The genus of a compact Riemann surface Σ is usually defined by

$$g(\Sigma) = \frac{1}{2} \operatorname{rank}_{\mathbb{Z}} H_1(\Sigma, \mathbb{Z})$$

in topology. On the other hand, in complex geometry, it is defined by

$$g(\Sigma) = \dim_{\mathbb{C}} H^1(\Sigma, \mathcal{O}).$$

These two definitions coincide, since we have an isomorphism $H^1(\Sigma, \mathcal{O}) \cong H^0(\Sigma, \Omega) = \Omega^1(\Sigma)$ by the Serre duality theorem, and the duality $\Omega^1(\Sigma) \cong H_1(\Sigma, \mathbb{R})$ as real vector spaces obtained by the Abel–Jacobi theorem.

1.5 Intersection Numbers and the Adjunction Formula

In this section, we review the definition of the intersection number of two divisors, the adjunction formula and its extension and the definition of the genus for a possibly singular complex curve. The contents here are basically due to Kodaira's paper [96], but the description becomes a bit longer, since we give detailed explanations about the way to deal with singularities.

Let V be a compact complex surface, E and F holomorphic line bundles over V, and C and D divisors on V. The *intersection number* of E and F is given by $(E \cdot F) = (c(E)c(F))(V)$. Using this definition, the *intersection number* of C and D is given by $(C \cdot D) = ([C] \cdot [D])$. Moreover, we write $(F \cdot D)$ for $(F \cdot [D]) = c(F)c([D])(V)$.

Then we have $(F \cdot D) = c(F)(D)$ and $(C \cdot D) = \sum_p I_p(C, D)$, where $I_p(C, D)$ is the intersection multiplicity of C and D over each $p \in C \cap D$. The *self-intersection number* $(C^2) = (C \cdot C)$ is identified with the Chern class of the normal bundle $N_{C/V}$ of $C \subset V$. Indeed, we have

$$(C^2) = c([C])(C) = c([C]|_C)(C) = c(N_{C/V})(C).$$

The last equality follows from the next theorem, called the *adjunction formula*.

Theorem 1.3 (Adjunction Formula) *Let W be a closed non-singular hypersurface in a compact complex manifold X, and N the normal bundle of $W \subset X$. Then, the following isomorphisms hold:*

$$N \cong [W]|_W, \quad K_W \cong (K_X)|_W \otimes [W].$$

Proof Let $\{U_\alpha\}$ be an atlas of X and $\{f_\alpha\}$ a family of local holomorphic functions defining the divisor W. Then the transition functions of the line bundle $[W]$ are given by $\{g_{\alpha\beta} = f_\alpha/f_\beta\}$. Derivating both sides of $f_\alpha = g_{\alpha\beta} f_\beta$, we obtain

$$df_\alpha = g_{\alpha\beta} df_\beta + f_\beta dg_{\alpha\beta}.$$

Hence, we have the equality $df_\alpha = g_{\alpha\beta} df_\beta$ on W. Each df_α defines a local holomorphic non-vanishing section of N^* over $U_\alpha \cap W$, and so, the family $\{df_\alpha\}$ forms a global non-vanishing section of $N^* \otimes [W]|_W$. Therefore, $N^* \cong [W]^{-1}|_W$, namely, $N \cong [W]|_W$.

Taking a local non-vanishing section ω_α of K_W and a local non-vanishing section df_α of N^* over $U_\alpha \cap W$, we obtain a local non-vanishing section $\omega_\alpha \wedge df_\alpha$ of $(K_X)|_W$. Therefore, we obtain $K_W \otimes N^* \cong (K_X)|_W$, namely, $K_W \cong (K_X)|_W \otimes [W]$. □

Now let C be a possibly reducible and singular curve in V and $\mu \colon \widetilde{C} \to C$ a resolution, that is, a holomorphic map μ from a non-singular curve $\widetilde{C}$ which is biholomorphic away from the preimage of the singular points of C. Then we can describe C and $\widetilde{C}$ as $C = \sum_\lambda C_\lambda$ and $\widetilde{C} = \sum_\lambda \widetilde{C}_\lambda$, where each C_λ is an irreducible component of C and $\widetilde{C}_\lambda$ is the non-singular model of C_λ. Let p be a singular point of C, whose inverse image is given by $\mu^{-1}(p) = \{q_1, \ldots, q_r\}$, and (x, y) local coordinates on V with the center p. Moreover, we take the local complex parameter t_i on $\widetilde{C}$ with the center q_i, and the minimal local equation $f_p(x, y)$ of C at p. For each singular point $p \in C$, we can associate the divisor

$$\mathfrak{c}_p = \sum_{i=1}^{r} m_i q_i$$

on $\widetilde{C}$, where m_i is the integer determined by

$$\| \mathrm{grad}(f_p)\,(\mu(t_i)) \| = |t_i|^{m_i} \|\mu'(t_i)\|.$$

(In fact, the integer $\deg(\mathfrak{c}_p) - r + 1$ coincides with the so-called *Milnor number* of f_p. See Remark 1.3.) Then we obtain the divisor

$$\mathfrak{c} = \sum_{p:\ \text{singular}} \mathfrak{c}_p,$$

which is called the *conductor* of C. For such a curve $C \subset V$, the adjunction formula can be written in the following form:

Theorem 1.4

$$K_{\widetilde{C}} \cong (K_V)|_{\widetilde{C}} \otimes [C] \otimes [\mathfrak{c}]^{-1}.$$

We don't give its proof here, but add some explanations about this isomorphism by considering the following two cases:

1. $V = \mathbb{C}^2$ and $C = \{(x, y) \in \mathbb{C}^2 \mid x^2 - y^3 = 0\}$.
2. $V = \mathbb{C}^2$ and $C = \{(x, y) \in \mathbb{C}^2 \mid xy = 0\}$.

The singularity at $(0, 0)$ defined by $x^2 - y^3 = 0$ is called an A_2-*singularity* or a *cusp*, and it appears in a *singular fiber of type* II in an elliptic surface (see Theorem 2.2). On the other hand, the second singular curve has a non-degenerate singularity at $(0, 0)$, and it is equivalent to the complex Morse singularity, which is defined by $x^2 + y^2 = 0$. This singularity is called an A_1-*singularity*, a *node* or an *ordinary double point*, which appears in a *singular fiber of type* I_1 in an elliptic surface. One might be worried that neither C nor $\mathbb{C}^2$ are compact, but what we would like to explain here is the reason that the term $[\mathfrak{c}]^{-1}$ arises. For this purpose, it suffices to focus on the local situation around the singularity.

Now we consider the case 1. First, by applying a quadratic transformation

$$(x, y) \mapsto \left(y, \; z = \frac{x}{y}\right),$$

we obtain the non-singular model $\widetilde{C} = \{(y, z) \in \mathbb{C}^2 \mid y = z^2\}$, where the resolution map is given by $\mu(y, z) = (yz, y)$. This can also be interpreted as follows. The non-singular model of C can be realized in $\mathbb{C}^3$ as the smooth curve

$$\hat{C} = \{(t^3, t^2, t) \in \mathbb{C}^3 \mid t \in \mathbb{C}\}$$

lying on the quadratic surface $\hat{V} = \{(x, y, z) \in \mathbb{C}^3 \mid x = yz\}$. By the projection $\pi_{12} : \mathbb{C}^3 \to \mathbb{C}^2; (x, y, z) \mapsto (x, y)$, it is mapped on the original singular curve C. On the other hand, another projection $\pi_{23} : \mathbb{C}^3 \to \mathbb{C}^2; (x, y, z) \mapsto (y, z)$ maps $\hat{V}$ to $\widetilde{V} = \mathbb{C}^2$, and $\hat{C}$ to $\widetilde{C}$ biholomorphically. Hence, it is natural to take t as a parameter of the smooth curve $\widetilde{C}$. Then, from the holomorphic forms $dx \wedge dy$ and $d(x^2 - y^3)$ on $V = \mathbb{C}^2$, we obtain the forms

$$d(yz) \wedge dy = -y\,dy \wedge dz \quad \text{and} \quad d(y^2 z^2 - y^3) = (2yz^2 - 3y^2)dy + 2y^2 z\,dz$$

on $\widetilde{V} = \mathbb{C}^2$. The restrictions of them to $\widetilde{C}$ are written as $-t^2 dy \wedge dz$ and $-t^4 dy + 2t^5 dz$, which give local sections of $K_V|_{\widetilde{C}}$ and $[C]^{-1}|_{\widetilde{C}}$, respectively. On the other hand, we have another 1-form

$$\frac{2t}{1 + 4t^2}dy + \frac{1}{1 + 4t^2}dz$$

along $\tilde{C}$, which corresponds to the natural section dt of $K_{\tilde{C}}$. Hence, we obtain the form

$$\left(\frac{2t}{1+4t^2}dy + \frac{1}{1+4t^2}dz\right) \wedge (-t^4 dy + 2t^5 dz) = t^4 dy \wedge dz$$

as a natural local section of $K_{\tilde{C}} \otimes [C]^{-1}$. Now we recall that in the non-singular case, the local sections of $K_C \otimes [C]^{-1}$ and $K_V|_C$ are non-vanishing. Taking this into account, we should treat the 2-forms $-t^2 dy \wedge dz$ and $t^4 dy \wedge dz$ as non-vanishing sections of

$$K_V|_{\tilde{C}} \otimes [0]^2 \quad \text{and} \quad K_{\tilde{C}} \otimes [C]^{-1} \otimes [0]^4,$$

respectively, where $0 \in \tilde{C} \subset \tilde{V} = \mathbb{C}^2$. Therefore, we obtain the isomorphism

$$K_{\tilde{C}} \otimes [0]^2 \cong K_V|_{\tilde{C}} \otimes [C].$$

Notice that the index 2 ($= 4 - 2$) coincides with the conductor of the cusp. This is verified by the following easy computations:

$$\mathrm{grad}(x^2 - y^3)|_{x=t^3, y=t^2} = (2t^3, -3t^4) = t^2(2t, -3t^2), \quad \frac{d}{dt}(t^3, t^2) = (3t^2, 2t).$$

Next, we consider the case 2. In this case, C consists of the two components $C_1 = \{y = 0\}$ and $C_2 = \{x = 0\}$, each of which is non-singular. We take their parameterizations by $C_1 = \{(t, 0) \mid t \in \mathbb{C}\}$ and $C_2 = \{(0, s) \mid s \in \mathbb{C}\}$. Hence, the non-singular models are given by $\tilde{C} = \tilde{C}_1 \cup \tilde{C}_2$, where $\tilde{C}_1 = \mathbb{C}$ and $\tilde{C}_2 = \mathbb{C}$. Then the resolution map μ consists of

$$\mu_1: \tilde{C}_1 \to C_1; t \mapsto (t, 0) \quad \text{and} \quad \mu_2: \tilde{C}_2 \to C_2; s \mapsto (0, s).$$

By a computation similar to that in the case 1, we obtain the 2-forms

$$t dx \wedge dy \quad \text{and} \quad -s dx \wedge dy$$

along $\tilde{C}_1$ and $\tilde{C}_2$, respectively. These 2-forms form a local non-vanishing section of $K_{\tilde{C}} \otimes [C]^{-1} \otimes [0_t] \otimes [0_s]$. Therefore, we obtain the isomorphism

$$K_{\tilde{C}} \otimes [0_t] \otimes [0_s] \cong K_V|_{\tilde{C}} \otimes [C].$$

Remark 1.3 Let $f(x, y)$ be a complex polynomial having an isolated critical point at $0 \in \mathbb{C}^2$. Then $f^{-1}(0)$ is an algebraic curve with an isolated singularity 0. We denote by r the number of irreducible components of it. On the other hand, $f^{-1}(c)$ ($c \neq 0, |c| \ll 1$) is a non-singular algebraic curve. Let μ be its first Betti number.

This is called the *Milnor number* of $(f, \mathbf{0})$. For the Milnor fibration, see Sect. 5.2. Then, the following holds in general:

$$\deg(c_0) = \mu + r - 1.$$

In particular, if $f^{-1}(0)$ is irreducible, we obtain the equality $\deg(c_0) = \mu$.

Proof of the Equality $\deg(c_0) = \mu$ First, we note that the gradient vector field

$$\operatorname{grad}(f) = (f_x, f_y) = \left(\frac{\partial f}{\partial x}, \frac{\partial f}{\partial y} \right)$$

is perpendicular to all the level sets of f with respect to the standard Hermitian metric on $\mathbb{C}^2$. In particular, it is perpendicular to the velocity vector $(c_1'(t), c_2'(t))$ along $f^{-1}(0)$. Thus, when $r = 1$, we obtain the equality

$$(-f_y, f_x)(c(t)) = t^{\deg(c_0)} (c_1'(t), c_2'(t)).$$

This shows the geometric meaning of $\deg(c_0)$. Pulling back $(-f_y, f_x)$ by the map $c \colon \mathbb{C} \to f^{-1}(0)$, we obtain a vector field X_f on the t-complex plane. Then, $\deg(c_0)$ is nothing but the rotation number of X_f along the unit circle $S^1 = \{t \in \mathbb{C} \mid |t| = 1\}$.

On the other hand, $(-f_y, f_x)$ also gives a non-vanishing vector field on any non-singular level set $f^{-1}(c)$. As a real surface, $f^{-1}(c)$ becomes a closed oriented surface $\Sigma_{\mu/2}$ of genus $\mu/2$ if we glue the 2-disk D^2 along the boundary. Let Y be any extension of the vector field $(-f_y, f_x)$ over $\Sigma_{\mu/2}$. Then, all the zeros of Y exist on D^2 and the sum of their indices are $\chi(\Sigma_{\mu/2}) = 2 - \mu$ by the Poincaré–Hopf theorem. By collapsing μ vanishing cycles of $f^{-1}(c)$, we obtain from $\Sigma_{\mu/2}$ a singular S^2 that is the union of $f^{-1}(0)$ and D^2. Now let us try to extend the vector field $Y|_{D^2}$ over $f^{-1}(0)$. Instead, this can be considered on the t-complex disk $\{t \in \mathbb{C} \mid |t| \le 1\}$. Recall that the rotation number of the vector field along the unit circle coincides with the obstruction to the extension as a non-vanishing vector field over the disk. Then, again by the Poincaré–Hopf theorem, the obstruction is

$$\chi(S^2) - (2 - \mu) = \mu$$

in our case. Therefore, the rotation number of X_f along $S^1 = \{t \in \mathbb{C} \mid |t| = 1\}$ is μ. This completes the proof of the equality $\deg(c_0) = \mu$. $\square$

For an irreducible projective curve $C \subset \mathbb{C}P^2$ of degree d, the following formula, called the Plücker formula, is known:

$$g(\tilde{C}) = \frac{(d-1)(d-2)}{2} - \frac{1}{2} \sum_{p:\,\text{singular}} c_p, \tag{1.1}$$

where c_p denotes the degree of the divisor $\mathfrak{c}_p$. However, $g(\widetilde{C})$ coincides with the topological genus of C, and does not reflect any information about singularities. Therefore, from the viewpoint of algebraic geometry, the integer

$$g(\widetilde{C}) + \frac{1}{2} \sum_{p:\,\text{singular}} c_p = \frac{(d-1)(d-2)}{2}$$

is more suitable for being called the genus of C.

On the other hand, it follows from Theorem 1.4 that

$$\sum_{\lambda} \left(2g(\widetilde{C}_\lambda) - 2\right) + \sum_{p:\,\text{singular}} c_p = C^2 + K_V \cdot C. \tag{1.2}$$

In particular, if C is irreducible and non-singular, we obtain

$$2g(C) - 2 = C^2 + K_V \cdot C \iff g(C) = \frac{1}{2}(C^2 + K_V \cdot C) + 1.$$

Given this equality, the following definition would be natural.

Definition 1.3 (Virtual Genus of a Line Bundle)   For a line bundle F on a compact complex surface V, its *virtual genus* $g'(F)$ is defined by

$$g'(F) = \frac{1}{2}(F^2 + K_V \cdot F) + 1.$$

We also define the *virtual genus of a divisor* D by $g'(D) = g'([D])$.

Then, the equality (1.2) can be rephrased as

$$\sum_{\lambda} \left(2g(\widetilde{C}_\lambda) - 2\right) + \sum_{p:\,\text{singular}} c_p = 2g'(C) - 2,$$

that is,

$$g'(C) = 1 + \sum_{\lambda} \left(g(\widetilde{C}_\lambda) - 1\right) + \frac{1}{2} \sum_{p:\,\text{singular}} c_p.$$

In particular, in the case where C is irreducible, we obtain

$$g'(C) = g(\widetilde{C}) + \frac{1}{2} \sum_{p:\,\text{singular}} c_p.$$

Moreover, if $C \subset \mathbb{C}P^2$ is a projective curve of degree d, the formula (1.1) implies that

$$g'(C) = \frac{(d-1)(d-2)}{2}.$$

Therefore, the virtual genus $g'(C)$ can be regarded as the algebro-geometric genus of C. Now we have given enough explanation about the adjunction formula and the genus of a complex curve.

1.6   The Riemann–Roch Theorem and its Extension

In this section, we explain about the Riemann–Roch theorem and its extension. Before that, we review the definition of the holomorphic Euler characteristic and the Serre duality theorem.

Definition 1.4 (Holomorphic Euler Characteristic)   For any holomorphic line bundle E on a compact complex manifold X, we define the *holomorphic Euler characteristic* $\chi(X, E)$ by

$$\chi(X, E) = \sum_{i=0}^{n} (-1)^i \dim H^i(X, O(E)).$$

In the case where X is a compact Riemann surface Σ of genus g and E is the trivial line bundle, we have $\chi(\Sigma, O) = 1-g$, since $H^0(\Sigma, O) = \mathbb{C}$ and $H^1(\Sigma, O) = \mathbb{C}^g$. This is the half of the topological Euler characteristic $\chi(\Sigma) = 2-2g$. In general, however, there is no relation between $\chi(X, E)$ and $\chi(X)$.

Theorem 1.5 (Serre Duality Theorem [163])   *Let E be a holomorphic vector bundle on a compact complex n-manifold X. Then the following isomorphism holds:*

$$H^q(X, \Omega^p(E)) \cong H^{n-q}(X, \Omega^{n-p}(E^*)).$$

Putting $p = 0$ in Theorem 1.5, we obtain the following.

Corollary 1.1 $H^q(X, O(E)) \cong H^{n-q}(X, O(K_X \otimes E^*))$.

The statement of the classical Riemann–Roch theorem is as follows.

Theorem 1.6 (Riemann–Roch)   *Let Σ be a compact Riemann surface of genus g and D a divisor on it. Then, the following equality holds:*

$$\dim H^0(\Sigma, O(D)) - \dim H^0(\Sigma, O(K_\Sigma \otimes [D]^{-1})) = \deg(D) + 1 - g.$$

In particular, we obtain

$$\dim H^0(\Sigma, O(D)) \geq \deg(D) + 1 - g,$$

which implies that for a divisor D with large positive degree, there exist a lot of holomorphic sections on the line bundle $[D]$. This is in contrast to the fact that the line bundle $[D]$ has no holomorphic section if $\deg D < 0$. Moreover, with respect to Corollary 1.1, we can rephrase the theorem as

$$\chi(\Sigma, [D]) - \chi(\Sigma, O) = \deg(D), \tag{1.3}$$

since $H^0(\Sigma, O(K_\Sigma \otimes [D]^{-1})) = H^1(\Sigma, O(D))$ and $\chi(\Sigma, O) = 1 - g$.

Proof of Theorem 1.6 Let p be any point on X. From the exact sequence of sheaves

$$0 \to O(D - p) \to O(D) \to O_p \to 0,$$

we obtain the exact sequence

$$0 \to H^0(\Sigma, O(D - p)) \to H^0(\Sigma, O(D)) \to \mathbb{C}$$
$$\to H^1(\Sigma, O(D - p)) \to H^1(\Sigma, O(D)) \to 0.$$

Therefore, we have

$$\chi(\Sigma, [D]) = \chi(\Sigma, [D - p]) + 1.$$

Then, by induction, we obtain Eq. (1.3).   □

Remark 1.4 For a compact Riemann surface Σ, we can show that the homomorphism Φ in Sect. 1.3 is an isomorphism. Let L be any line bundle over Σ. Then, by the same argument as above, we have

$$\chi(\Sigma, L \otimes [np]) = \chi(\Sigma, L) + n$$

for any $p \in \Sigma$ and $n \in \mathbb{Z}$. Hence, for a sufficiently large positive n, $\chi(\Sigma, L \otimes [np]) > 0$, and in particular,

$$\dim H^0(\Sigma, O(L \otimes [np])) > 0.$$

Therefore, the line bundle $L \otimes [np]$ has a global holomorphic section and we obtain a divisor D as the zero set of it. Then, $[D] = L \otimes [np]$, and hence, $L = [D - np]$.

Theorem 1.7 (Kodaira [95])   *Let V be a compact complex surface. For any holomorphic line bundle F and any divisor D on V, the following equality holds:*

$$\chi(V, F) = \chi(V, F \otimes [D]^{-1}) + (F \cdot D) - g'(D) + 1.$$

This can be considered as the surface version of the Riemann–Roch theorem, since we can rephrase it in the following form:

$$\chi(V, F) - \chi(V, F \otimes [D]^{-1}) = \chi(D, F|_D) \tag{1.4}$$

which is proven by the long exact sequence obtained from the sheaf exact sequence

$$0 \to O(F \otimes [D]^{-1}) \to O(F) \to O(F|_D) \to 0.$$

Theorems 1.6 and 1.7 have been generalized by Hirzebruch [71] to the theorem for compact complex manifolds of any dimension, which is now called the *Hirzebruch–Riemann–Roch theorem*.

1.7 Symplectic Manifolds and Kähler Manifolds

In this section, we give a minimal review of symplectic manifolds and Kähler manifolds. Good references are [5, 123] for symplectic geometry, and [9, 175] for Kähler geometry.

Definition 1.5 (Symplectic Structures, Symplectic Manifolds) A closed 2-form ω on a real $2n$-dimensional manifold M^{2n} is called a *symplectic form* or a *symplectic structure* if ω is non-degenerate, that is, the n-th wedge power ω^n is a positive volume form on M^{2n}. In this case, we say that (M^{2n}, ω) is a *symplectic manifold*.

Example 1.4 We give the following three fundamental examples.

1. A real 2-dimensional orientable surface admits a symplectic structure, since in this dimension ($n = 1$), a symplectic form is nothing but a positive area form.
2. Let $(x_1, y_1, \ldots, x_n, y_n)$ be the Cartesian coordinates on $\mathbb{R}^{2n}$. Then the 2-form

$$\omega_0 = \sum_{j=1}^{n} dx_j \wedge dy_j$$

is a symplectic form. This is called the *standard symplectic form* and $(\mathbb{R}^{2n}, \omega_0)$ is called the *standard symplectic structure* on $\mathbb{R}^{2n}$.
3. Let M^n be an n-manifold and $(p_1, \ldots, p_n, q_1, \ldots, q_n)$ local coordinates on the cotangent bundle T^*M^n, where $(p_1, \ldots, p_n)$ and $(q_1, \ldots, q_n)$ are the coordinates on a fiber and the base manifold, respectively. Since the 1-form $\sum_{j=1}^{n} p_j \wedge dq_j$ does not depend on the choice of local coordinates $(q_1, \ldots, q_n)$ on M^n, we obtain a global 1-form on T^*M^n having this local description. Taking the derivation of the 1-form, we obtain the canonical symplectic form on T^*M^n

$$\omega = \sum_{j=1}^{n} dp_j \wedge dq_j.$$

Notice that the standard symplectic structure $(\mathbb{R}^{2n}, \omega_0)$ can be seen as the cotangent bundle of $\mathbb{R}^n$. Let (M_1, ω_1) and (M_2, ω_2) be symplectic manifolds. A diffeomorphism $f: M_1 \to M_2$ is called a *symplectomorphism* if $f^*\omega_2 = \omega_1$. In symplectic geometry, the following theorem is fundamental.

Theorem 1.8 (Darboux) *Any symplectic manifold (M^{2n}, ω) is locally symplectomorphic to an open set of $(\mathbb{R}^{2n}, \omega_0)$.*

In other words, a symplectic manifold is the resultant manifold of a gluing of open sets of $(\mathbb{R}^{2n}, \omega_0)$ by symplectomorphisms. A sympletic manifold (M^{2n}, ω) is said to be *exact* if ω is an exact 2-form, and such a symplectic form ω is called an *exact symplectic form*. The cotangent bundle (T^*M^n, ω) is a typical exact symplectic manifold. Notice that if M^{2n} is a closed manifold, then (M^{2n}, ω) is never exact, since the symplectic form ω represents a non-trivial element of $H^2_{DR}(M^{2n})$.

Let L^n be a submanifold of (M^{2n}, ω) and $i: L^n \to M^{2n}$ the inclusion map. Then L^n is called a *Lagrangian submanifold* if $i^*\omega = 0$ holds on L^n. Typical examples are the Clifford torus $T^n = (S^1)^n \subset \mathbb{C}^n$ and the graph of a closed 1-form on M^n, which is a submanifold of the cotangent bundle T^*M^n. In particular, the zero section of the cotangent bundle $M^n \subset T^*M^n$ is a Lagrangian submanifold.

Theorem 1.9 (Weinstein Tubular Neighborhood Theorem) *Let L^n be a Lagrangian submanifold of (M^{2n}, ω). Then there exists a tubular neighborhood $N(L^n) \subset M^{2n}$ of L^n that is symplectomorphic to a neighborhood of the zero section $L^n \subset T^*L^n$.*

Lagrangian submanifolds are important objects in symplectic geometry. For example, the Arnol'd–Liouville theorem [4, 112] says that an integrable Hamiltonian system is nothing but a Lagrangian torus fibration if all the orbits are compact. It also plays a central role in Lagrangian Floer theory and J-holomorphic curve techniques, which originated from Gromov's non-squeezing theorem [66], and leads to the solution to the Arnol'd conjecture (Fukaya and Ono [52], Conley and Zehnder [26], Floer [46–48], Hofer [73], Ono [148]). For Lagrangian Floer theory and J-holomorphic curves, see [122].

Next, we review Kähler manifolds. Let X be a complex manifold and J the complex structure on it. Moreover, we assume that X, as an even-dimensional real manifold, admits a symplectic structure ω. We say that ω and J are *compatible* if the following two conditions are satisfied:

1. $\omega(Ju, Jv) = \omega(u, v)$,
2. $\omega(u, Ju) > 0 \quad (u \neq 0)$.

Notice that the compatibility is equivalent to the condition that $g(u, v) := \omega(u, Jv)$ is a Riemannian metric.

Definition 1.6 (Kähler Manifolds) A complex manifold (X, J) is said to be *Kähler* if there exists a symplectic form ω compatible with J.

Definition 1.7 (Hermitian Metrics, Kähler Metrics) Let X be a complex manifold and J the complex structure on it . A Riemannian metric g on X is called a *Hermitian metric* if it is J-invariant. For a Hermitian metric g, the 2-form ω given by $\omega(u, v) = g(Ju, v)$ is J-invariant and non-degenerate. If the 2-form ω is closed, in addition, g and ω are called a *Kähler metric* and a *Kähler form*, respectively.

Remark 1.5 Hermitian metrics always exist on a complex manifold. Indeed, from any Riemannian metric g on a complex manifold (X, J), we obtain a Hermitian metric $\tilde{g}$ by

$$\tilde{g}(u, v) = \frac{g(u, v) + g(Ju, Jv)}{2}.$$

A Kähler form is nothing but a symplectic form ω compatible with the complex structure J. If ω is a Kähler form, then $g(u, v) = \omega(u, Jv)$ is a Kähler metric. The most fundamental example of Kähler manifold is $\mathbb{C}^n$, where the canonical Kähler form ω_0 is given by

$$\omega_0 = \frac{i}{2}\partial\bar{\partial}\left(\sum_{j=1}^n |z_j|^2\right) = \frac{i}{2}\sum_{j=1}^n dz_j \wedge d\bar{z}_j.$$

This coincides with the standard symplectic form on $\mathbb{R}^{2n}$. The next fundamental example is $\mathbb{C}P^n$. The canonical Kähler form on $\mathbb{C}P^n$ is given as follows:

$$\omega = \frac{i}{2}\partial\bar{\partial}\log\left(1 + \sum_{j=1}^n |z_j|^2\right).$$

The corresponding Riemannian metric g on $\mathbb{C}P^n$ defined by $g(u, v) = \omega(u, Jv)$ is called the *Fubini–Study metric*.

Example 1.5 The following are typical examples of Kähler manifolds.

1. Riemann surfaces. Since the real dimension is 2 ($n = 1$), any 2-form is closed. Hence, every Hermitian metric on a Riemann surface is a Kähler metric.
2. Stein manifolds (see Sect. 5.2). Since a Stein manifold is a complex submanifold of some $\mathbb{C}^n$, the restriction of ω_0 gives the canonical Kähler form. More generally, a strongly pseudoconvex manifold is Kähler, since the 2-form $i\partial\bar{\partial}\varphi$ is a Kähler form for a strictly plurisubharmonic function φ.
3. Projective manifolds. Since a projective manifold is a complex submanifold of some $\mathbb{C}P^n$, the restriction of the Fubini–Study metric gives the canonical Kähler metric.
4. Complex tori. Let Λ be a lattice in $\mathbb{C}^n$. Then the quotient $\mathbb{C}^n/\Lambda$ is a compact complex n-manifold. This is called an *n-dimensional complex torus*. It is Kähler, since the canonical Kähler form ω_0 on $\mathbb{C}^n$ is translation invariant. A 2-dimensional complex torus is a fundamental example of compact Kähler surface.

5. K3 surfaces. A simply-connected compact complex surface with trivial canonical bundle is called a *K3 surface*. A smooth projective quintic surface is a typical example. It was shown by Kodaira that all the K3 surfaces are diffeomorphic to each other. Moreover, it is well-known that every K3 surface is Kähler. This is a highly non-trivial fact, which is due to Siu (Theorem 2.7).

There are still some important classes of compact Kähler surfaces, called *rational surfaces* and *ruled surfaces*. A rational surface is usually defined as a compact complex surface birational to $\mathbb{C}P^2$. It is well-known that a minimal rational surface is either $\mathbb{C}P^2$ or a Hirzebruch surface F_n with $n \geq 0$ and $n \neq 1$. Thus, a general rational surface is such a surface or its blow-up. In this book, we adopt this property as the definition of rational surfaces, since we don't want to define the birationality here. A compact complex surface is called a *ruled surface* of genus g if it is the total space of a holomorphic $\mathbb{C}P^1$-bundle over a smooth complex curve of genus g. Notice that a ruled surface of genus 0 is exactly a Hirzebruch surface. Both rational surfaces and ruled surfaces are all projective, and thus, they are Kähler surfaces.

On the other hand, a complex manifold is said to be *non-Kähler* if it admits no Kähler form. Hopf surfaces and Calabi–Eckmann manifolds (see Sects. 2.3 and 4.2) are obviously non-Kähler complex manifolds, since they have trivial second de Rham cohomology groups. Kodaira surfaces are also important examples of non-Kähler complex surfaces. In this case, the non-Kählerness is proven by the following fact.

Theorem 1.10 *Any odd-dimensional Betti number of a compact Kähler manifold is even.*

This is a consequence of the Hodge theory (see [9, 65, 175]), and plays an important role in Sect. 2.3.

1.8   Algebraic Dimension and Kodaira Dimension

Let K be a field. A field extension L/K is called an *algebraic extension* if L is a field obtained from K by adding some roots of polynomials with coefficients in K. A field extension that is not algebraic is called a *transcendental extension*. For example, $\mathbb{Q}(\sqrt{2}) = \mathbb{Q}[x]/(x^2 - 2)$ and $\mathbb{C} = \mathbb{R}[x]/(x^2 + 1)$ are algebraic extensions over $\mathbb{Q}$ and $\mathbb{R}$, respectively. On the other hand, $\mathbb{C}(x)$ is a transcendental extension of $\mathbb{C}$, whose transcendental degree is 1.

Let V be a compact complex manifold of complex dimension n, and $\mathcal{M}(V)$ the field of all meromorphic functions on V. For this space, the following result is known.

Theorem 1.11 (Siegel [164])   *The field $\mathcal{M}(V)$ is a finite algebraic function field over $\mathbb{C}$ whose transcendental degree is not greater than n.*

The transcendental degree of $M(V)$ is called the *algebraic dimension* of V and denoted by $a(V)$. Then, $M(V)$ is a finite algebraic extension of $\mathbb{C}(x_1, \ldots, x_k)$, where $k = a(V)$ and $x_1, \ldots, x_k$ are some meromorphic functions on V. Hence, any meromorphic function is obtained from $x_1, \ldots, x_k$ algebraically, that is, $M(V)$ can be described in the form

$$M(V) = \mathbb{C}(x_1, \ldots, x_k, y_1, \ldots, y_l)/I,$$

where I is an ideal generated by a finite number of polynomials of $x_1, \ldots, x_k$, $y_1, \ldots, y_l$. Therefore, there exists an algebraic variety V^* of dimension $a(V)$ in a projective space such that $M(V) \cong M(V^*)$. Combining with Hironaka's resolution theorem, we obtain the following as a corollary to Theorem 1.11.

Corollary 1.2 *For any compact complex manifold V, there exists a non-singular projective algebraic manifold V^* of dimension $a(V)$ such that $M(V) \cong M(V^*)$.*

For a compact complex manifold V, the *d-th plurigenus P_d* is defined by

$$P_d = \dim_{\mathbb{C}} H^0(V, K_V^d),$$

where K_V is the canonical line bundle of V.

Definition 1.8 (Kodaira Dimension) The Kodaira dimension $\kappa(V)$ of a compact complex manifold V is defined as follows:

$$\kappa(V) = \begin{cases} -\infty & \text{(if } P_d = 0 \text{ for any } d > 0), \\ \min\left\{ k \in \mathbb{Z}_{\geq 0} \mid \{d^{-k} P_d\}_{d=1}^{\infty} \text{ is bounded} \right\} & \text{(otherwise).} \end{cases}$$

It is known that if $\kappa(V) \neq -\infty$, then $0 \leq \kappa(V) \leq \dim(V)$.

Now let us consider the case where V is a compact complex surface. Then, its Kodaira dimension satisfies $\kappa(V) = -\infty, 0, 1$ or 2. If the Kodaira dimension is maximal, namely, $\kappa(V) = 2$, V is called a *surface of general type*. Notice that if the canonical line bundle K_V is trivial, then $P_d = 1$ for any $d > 0$, and hence, $\kappa(V) = 0$. In particular, the Kodaira dimension of a K3 surface and a complex torus is 0.

Chapter 2
Compact Complex Surfaces

In this chapter, we review the theory of compact complex surfaces. Especially, we focus on Kodaira's elliptic surface theory in Sect. 2.2 and various examples of compact non-Kähler surfaces in Sect. 2.3.

2.1 Kodaira's Original Classification

Nowadays "the classification of compact complex surfaces" usually means the classification into the following 10 classes using the Kodaira dimension κ, and the first and second Betti numbers b_1, b_2. This is called the Enriques–Kodaira classification.

1. rational surfaces ($\kappa = -\infty$, $b_1 = 0$),
2. ruled surfaces ($\kappa = -\infty$, $b_1 = 2g$, $g \geq 1$),
3. surfaces of class VII ($\kappa = -\infty$, $b_1 = 1$),
4. complex tori ($\kappa = 0$, $b_1 = 4$),
5. K3 surfaces ($\kappa = 0$, $b_1 = 0$, $b_2 = 22$),
6. Enriques surfaces ($\kappa = 0$, $b_1 = 0$, $b_2 = 10$),
7. bi-elliptic surfaces ($\kappa = 0$, $b_1 = 2$),
8. Kodaira surfaces ($\kappa = 0$, $b_1 = 3$ if primary, $b_1 = 1$ if secondary),
9. properly elliptic surfaces ($\kappa = 1$),
10. algebraic surfaces of general type ($\kappa = 2$).

On the other hand, the original classification given by Kodaira [102] is as follows.

(I) $\mathbb{C}P^2$ or ruled surfaces,
(II) K3 surfaces,
(III) complex tori,
(IV) Kähler elliptic surfaces,

© The Editor(s) (if applicable) and The Author(s), under exclusive license
to Springer Nature Singapore Pte Ltd. 2025
N. Kasuya, *Non-Kähler Complex Surfaces and Strongly Pseudoconcave Surfaces*,
SpringerBriefs in Mathematics, https://doi.org/10.1007/978-981-96-3002-8_2

(V) algebraic surfaces of general type,
(VI) non-Kähler elliptic surfaces with b_1 odd and $b_1 \geq 3$,
(VII) non-Kähler surfaces with $b_1 = 1$.

Here, the classification is done up to blow-ups and blow-downs by the pluri-genera, the first Betti number, the square of the first Chern class, and the canonical line bundle. Moreover, the algebraic dimension plays an important role, though it is not explicitly written in the table (Theorem 55 in [102]). We note here that the definition of surfaces of class VII is different for the two classifications. In the former classification, surfaces of class VII are those with $\kappa = -\infty$ and $b_1 = 1$, while in the latter one, all the surfaces with $b_1 = 1$ are of class VII. Therefore, secondary Kodaira surfaces are of class VII in the latter classification.

In this book, we adopt the original classification into the 7 classes, since we will focus on Kodaira's elliptic surface theory. Here we recall the definition of an elliptic surface.

Definition 2.1 An *elliptic fibration* on a complex surface V is a surjective holomorphic map $\Phi \colon V \to C$ with a general fiber $\Phi^{-1}(t)$ a smooth elliptic curve and C a smooth compact complex curve. An *elliptic surface* is a surface with an elliptic fibration on it.

Now we are going to survey the elliptic surface theory according to the series of papers by Kodaira in Sect. 2.2, and then in Sect. 2.3, after reviewing the theorem of Kodaira, Miyaoka and Siu, we address the surfaces of class VII.

2.2 Kodaira's Elliptic Surface Theory

In what follows, V denotes a compact complex surface. Then the algebraic dimension $a(V)$ is either 2, 1 or 0. In the series of papers by Kodaira, the subject is the classification of compact complex surfaces by the algebraic dimension, whose conclusion can be summarized as follows.

Theorem 2.1 (Kodaira [96–98]) *A compact complex surface V is classified by its algebraic dimension $a(V)$ as follows:*

(1) If $a(V) = 2$, then V is a projective algebraic surface.
(2) If $a(V) = 1$, then V is an elliptic surface over a projective algebraic curve.
(3) If $a(V) = 0$, then V is any of the following three types:

   (a) If $b_1 = 4$, then V is a complex torus.
   (b) If $b_1 = 1$, then V is of class VII.
   (c) If $b_1 = 0$, then V is a K3 surface.

In this section, we first outline the proofs of Theorem 2.1 (1) and (2), then review the theory of singular fibers and multiple fibers of elliptic surfaces. In those arguments, we will use many of the tools prepared in the previous chapter. Especially, the two theorems, the result of Siegel (Theorem 1.11) and the extension of the Riemann–Roch Theorem (Theorem 1.7), play crucial roles.

Now let us review the outline of the proof of Theorem 2.1 (1).

Outline of the Proof of Theorem 2.1 (1) By Theorem 1.11, there exists a non-singular algebraic surface $V^* \subset \mathbb{C}P^d$ such that $\mathcal{M}(V) \cong \mathcal{M}(V^*)$. Let $x_1, \ldots, x_d$ be the meromorphic functions given by $x_k = \frac{X_k}{X_0}$, where $[X_0 : X_1 : \cdots : X_d]$ are the homogeneous coordinate functions of $V^* \subset \mathbb{C}P^d$. Pulling them back through the isomorphism $\mathcal{M}(V) \cong \mathcal{M}(V^*)$, we obtain meromorphic functions $y_1, \ldots, y_d$ on V. Now we consider the natural meromorphic map

$$\Phi \colon V \to \mathbb{C}P^d \colon z \mapsto [1 \colon y_1(z) \colon \cdots \colon y_d(z)].$$

Let $C_\lambda^* \ (\lambda = [\lambda_0 \colon \lambda \colon \cdots \colon \lambda_d] \in \mathbb{C}P^d)$ be the hyperplane section of V^* defined by

$$\lambda_0 X_0 + \lambda_1 X_1 + \cdots + \lambda_d X_d = 0,$$

and $C_\lambda = \Phi^{-1}(C_\lambda^*)$. Then Φ is holomorphic except at the finite base points of the linear system $\{C_\lambda\}$. Since the multiplicity of each base point can be strictly decreased by a blow-up, these base points can be resolved by a finite sequence of blow-ups. Thus we obtain a holomorphic map $\widetilde{\Phi} \colon \widetilde{V} \to V^*$. Since the inverse map $\widetilde{\Phi}^{-1}$ is locally biholomorphic except at fundamental points and branch loci, $\widetilde{\Phi}$ restricted there is a covering map. Let m be the degree of the covering. Take a hypersurface section D_k^* of V^* of order k, and put $D_k = \widetilde{\Phi}^{-1}(D_k^*)$. The isomorphism $\mathcal{M}(\widetilde{V}) \cong \mathcal{M}(V^*)$ implies

$$\dim H^0(\widetilde{V}, O(D_k)) = \dim H^0(V^*, O(D_k^*)).$$

By using Theorem 1.7, we can estimate this value with respect to k. In fact the right-hand side is a quadratic function of k, and the left-hand side is bounded from below by another quadratic function of k, whose top terms are $(D_k^2) = (D_1^2)k^2$ and $m(D_k^{*2}) = m(D_1^{*2})k^2$, respectively. Therefore, we have $m = 1$, which means that $\widetilde{\Phi}$ is a biholomorphism except at finite fundamental points, and there is no branch locus.

Each fundamental point can be resolved again by a finite sequence of one point blow-ups. Thus we obtain an algebraic projective surface bihilomorphic to $\tilde{V}$ by blowing up V^* at these fundamental points. Since $\tilde{V}$ is an algebraic projective surface, so is V by Castelnuovo's theorem.   □

Based on Theorem 2.1 (1), we obtain the following result.

Proposition 2.1 (Theorem 3.3 in [96])   *If a compact complex surface V contains a positively embedded curve C, then V is a projective algebraic surface.*

Proof   For simplicity, we only prove the case where C is non-singular. Let $F = [C]$. From the short exact sequence of sheaves

$$0 \to O(F^{m-1}) \to O(F^m) \to O(F^m|_C) \to 0,$$

we obtain the exact sequence of cohomologies

$$
\begin{aligned}
0 &\longrightarrow &\Gamma(F^{m-1}) &\longrightarrow &\Gamma(F^m) &\longrightarrow &H^0(C, F^m|_C) \\
&\longrightarrow &H^1(V, F^{m-1}) &\longrightarrow &H^1(V, F^m) &\longrightarrow &H^1(C, F^m|_C) \\
&\longrightarrow &H^2(V, F^{m-1}) &\longrightarrow &H^2(V, F^m) &\longrightarrow &H^2(C, F^m|_C) = 0.
\end{aligned}
$$

Then, for a sufficiently large positive integer m, the short exact sequence

$$0 \to \Gamma(F^{m-1}) \to \Gamma(F^m) \to H^0(C, F^m|_C) \to 0$$

holds for the following reason. First, by the Serre duality theorem, we have

$$H^1(C, F^k|_C) = H^0(C, K_C \otimes F^{-k}).$$

Since $(C^2) \geq 1$, the degree of the line bundle $K_C \otimes F^{-k}$ is negative with k large. Hence, we have $H^0(C, K_C \otimes F^{-k}) = 0$, which implies the surjectivity of the map $H^1(V, F^{k-1}) \to H^1(V, F^k)$. Now we set $d_i = \dim H^1(V, F^i)$. Since the sequence $\{d_i\}_{i=k-1}^{\infty}$ is decreasing, there exists an integer m_0 such that d_i is constant for $i \geq m_0$. This means that for $m \geq m_0$, the surjection $H^1(V, F^{m-1}) \to H^1(V, F^m)$ becomes an isomorphism and the desired exact sequence holds.

Since $\deg(F^m|_C) = m(C^2) \geq m$, it follows from the Riemann–Roch theorem that

$$\dim H^0(C, F^m|_C) \to +\infty \quad (m \to +\infty).$$

Combined with the above short exact sequence, this also implies that

$$\dim \Gamma(F^m) \to +\infty \quad (m \to +\infty).$$

Hence there exists an integer m_1 such that if $m \geq m_1$, then

$$\dim H^0(C, F^m|_C) \geq 2, \quad \dim \Gamma(F^{m-1}) \geq 1.$$

Pulling back a nontrivial element of $H^0(C, F^m|_C)$, we obtain an element $\varphi \in \Gamma(F^m)$ such that $\varphi|_C \neq 0$. Though the divisor $D = (\varphi)$ does not contain C as a component, the corresponding line bundle is isomorphic to $[mC]$. Hence, $[D] = F^m$, $[D - C] = F^{m-1}$. Since $\dim H^0(C, F^m|_C) \geq 2$, there exists a meromorphic function x on V that has D as a pole, and is non-constant over C. On the other hand, by $\dim \Gamma(F^{m-1}) \geq 1$, there exists a non-constant meromorphic function y on V that is identically 0 over C. Since the two functions x and y are algebraically independent, V is a projective surface by Theorem 2.1 (1).

Even in the case where C is a singular curve, the proof is basically the same, though a little more careful argument is needed.   □

Now we outline the proof of Theorem 2.1 (2). The crucial part is the following proposition.

Proposition 2.2 (Theorem 4.1 in [96])   *Let V be a compact complex surface V with $a(V) = 1$. Then V admits a holomorphic fibration over a non-singular projective curve.*

Proof   Since the algebraic dimension of V is 1, by Theorem 1.11, there exists a non-singular projective curve C such that $\mathcal{M}(V) \cong \mathcal{M}(C)$. Hence, the meromorphic function X_k/X_0 on C obtained as a coordinate function of $C \subset \mathbb{C}P^n$ induces a meromorphic function x_k on V ($k = 1, \ldots, n$). For a sufficiently fine cover $\{U^\alpha\}$ of V, the function can be described as $x_k = \varphi_k^\alpha/\varphi_0^\alpha$ by holomorphic functions φ_k^α on U^α, and hence, we obtain the map

$$\Phi^\alpha : U^\alpha \to \mathbb{C}P^n; z \mapsto [\varphi_0^\alpha(z) : \varphi_1^\alpha(z) : \cdots : \varphi_n^\alpha(z)].$$

The maps Φ^α and Φ^β coincide on the intersection $U_\alpha \cap U_\beta$, so $\{\Phi^\alpha\}$ constitute a single map $\Phi : V \to \mathbb{C}P^n$. This is in fact a holomorphic map from V to C. Indeed, the image of Φ is contained in C by the isomorphism $\mathcal{M}(V) \cong \mathcal{M}(C)$, and the set of base points

$$B = \left\{ z \in V \mid \varphi_0^\alpha(z) = \cdots = \varphi_n^\alpha(z) = 0 \right\}$$

is empty by the following argument. First, let

$$S = \{z \in V \mid \varphi_\mu^\alpha(z)\partial_i\varphi_\tau^\alpha(z) - \varphi_\tau^\alpha(z)\partial_i\varphi_\mu^\alpha(z) = 0 \text{ for all } \mu, \tau, i, \alpha\},$$

then S is a subvariety of V and

$$\partial_i(\varphi_\mu^\alpha/\varphi_\tau^\alpha) = \frac{\varphi_\mu^\alpha\partial_i\varphi_\tau^\alpha - \varphi_\tau^\alpha\partial_i\varphi_\mu^\alpha}{(\varphi_\tau^\alpha)^2} = 0$$

holds on $S - B$. Therefore Φ is constant on each connected component of $S - B$ and thus $\Phi(S - B)$ is a finite set. Next, we set $D(u) = \{z \in V \mid u_\tau\varphi_\mu^\alpha(z) - u_\mu\varphi_\tau^\alpha(z) = 0\}$ for any $u \in C - \Phi(S - B)$. Then $D(u)$ is a curve containing B and non-singular outside B. Indeed, since each $z \in D(u) - B$ is not an element of S, there exist i, τ, μ such that

$$\partial_i(u_\tau\varphi_\mu^\alpha - u_\mu\varphi_\tau^\alpha) = u_\tau\partial_i\varphi_\mu^\alpha - u_\mu\partial_i\varphi_\tau^\alpha \neq 0.$$

Now suppose B were not empty. Then for a base point $b \in B$, there exist distinct two points $u, u' \in C$ such that $D_1(u)$ and $D_1(u')$ are homologous in $H_2(V; \mathbb{Z})$, where $D_1(u)$ is the connected component of $D(u)$ containing b. Then we have

$$D_1(u)^2 = D_1(u)D_1(u') \geq 1,$$

since the two algebraic curves $D_1(u)$ and $D_1(u')$ intersects at the point b. Hence, by Proposition 2.1, V must be projective, which contradicts the hypothesis that the algebraic dimension is 1. Therefore, Φ has no base point and is a holomorphic fibration from V to C. $\square$

Finally, Kodaira showed that a general fiber of Φ is irreducible with genus 1 by a smart argument on intersection numbers of divisors using Theorem 1.7 and some lemma (Lemma 4.1 in [97]). This completes the proof of Theorem 2.1 (2).

Now let V be an elliptic surface and $\pi : V \to C$ an elliptic fibration on it. A fiber $\pi^{-1}(a)$ is called a *singular fiber* if a is a critical value of π. A singular fiber can be written in the form $\pi^{-1}(a) = \sum_{i=1}^{k} n_i D_i$ as a divisor. The greatest common divisor of n_i $(i = 1, \ldots, k)$ is called the *multiplicity* of $\pi^{-1}(a)$. If it is not equal to 1, then the fiber $\pi^{-1}(a)$ is called a *multiple fiber*. In Kodaira's elliptic surface theory, the classification of possible singular fibers and multiple fibers of elliptic surfaces is the principal result. They are classified as follows.

Theorem 2.2 (Theorem 6.2 in [97]) *A singular fiber of a minimal elliptic surface is any of the following types:*

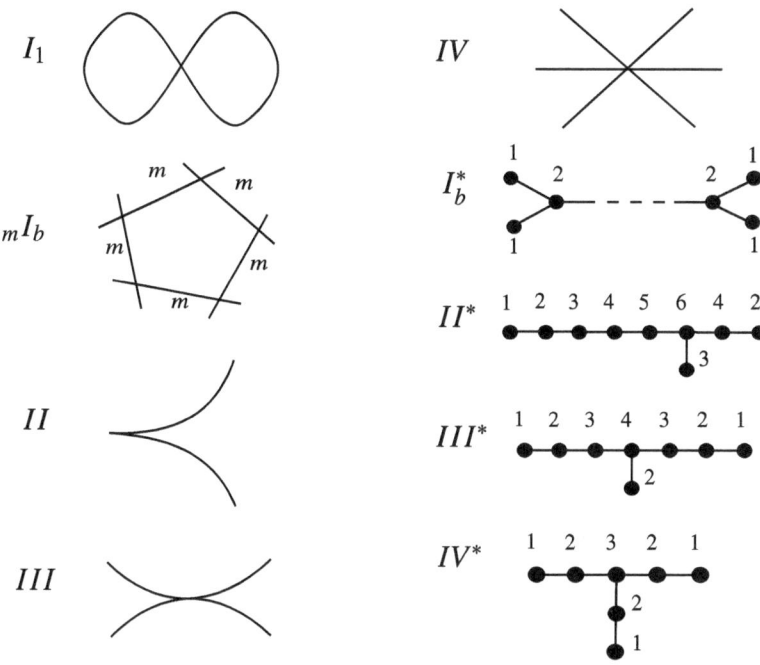

Here we have to explain what these diagrams mean. In the diagrams for I_1, $_mI_b$, II, III and IV, each line or curve represents an irreducible component of the divisor, which is a rational curve, and the diagram describes how they intersect each other. More concretely, I_1, $_mI_b$, II, III and IV denote a non-singular rational curve with one node, a cycle of b non-singular rational curves of multiplicity m, a singular rational curve with one cusp, two non-singular rational curves meeting at one point with intersection multiplicity 2, and three rational curves transversely intersecting at one point, respectively. The other diagrams are dual graphs of the divisors. Each vertex represents a non-singular rational curve and two vertices are connected by an edge if and only if the corresponding two rational curves transversely intersect at one point. On each vertex, there is written the coefficient of each component in the divisor representing the singular fiber. In the case of an irreducible singular fiber, namely, I_1, II, $_mI_0$ and $_mI_1$, the self-intersection number of the divisor is 0. In all the other cases, each non-singular rational curve has a self-intersection number -2.

Outline of the Proof Let V be a minimal elliptic surface and $\Phi\colon V \to S$ an elliptic fibration on it. We take a critical value a of Φ. Then the singular fiber $C(a) = \Phi^{-1}(a)$ can be described as

$$C(a) = \sum_{i=1}^{k} n_i D_i,$$

where each D_i is an irreducible curve. The divisor a on S is linearly equivalent to a divisor $\sum_{j=1}^{k} m_j u_j$ of degree 1, where each u_j is a regular value of Φ. Hence $C(a)$ is linearly equivalent to $\sum_{j=1}^{k} m_j C(u_j)$, while each $C(u_j)$ is homotopic to a regular fiber $C(u)$. Therefore, the singular fiber $C(a)$ is homologous to one regular fiber $C(u)$ as a 2-cycle, and we obtain

$$D_i \cdot C(u) = D_i \cdot C(a) = \sum_{j=1}^{k} n_j (D_i \cdot D_j) = 0. \tag{2.1}$$

Moreover, we have $g(C(u)) = \frac{1}{2}(C(u)^2 + K_V \cdot C(u)) + 1 = 1$ and $C(u)^2 = 0$. Hence, $K_V \cdot C(u) = 0$. Since $K_V \cdot D = 2g'(D) - 2 - (D^2)$ holds for any divisor D, we obtain

$$\sum_{i=1}^{k} n_i (2g'(D_i) - 2 - (D_i^2)) = 0. \tag{2.2}$$

The theorem can be proven by elementary arguments using Eqs. (2.1) and (2.2).

1. The case $k = 1$. We have $D_1^2 = 0$ and $g(D_1) = 1$. If D_1 contains no singularity, it is a non-singular elliptic curve (I_0). If it has a singularity, it follows from the formula $g'(D_1) = g(\tilde{D}_1) + \frac{1}{2} \sum c_p$ that $g'(\tilde{D}_1) = 0$ and $c_p = 2$. Therefore, D_1 is either a rational curve with a node (I_1), or a rational curve with a cusp (II).
2. The case $k > 1$. Since $n_i(D_i^2) = -\sum_{j \neq i} n_j (D_i \cdot D_j) \leq -1$, we have $(D_i^2) \leq -1$. If $g'(D_i) = 0$, then $(D_i^2) \leq -2$ by the minimality assumption on V. In this case,

$$2g'(D_i) - 2 - (D_i^2) = -2 - (D_i^2) \geq 0.$$

On the other hand, if $g'(D_i) \geq 1$ for some i, then we have

$$2g'(D_i) - 2 - (D_i^2) \geq -(D_i^2) > 0.$$

Hence, by the equality

$$\sum_{i=0}^{k} n_i (2g'(D_i) - 2 - (D_i^2)) = 0,$$

it must be that $g'(D_i) = 0$ and $(D_i^2) = -2$ for all i. Therefore,

$$2n_i = \sum_{j \neq i} n_j (D_i \cdot D_j).$$

Then, a careful argument about integer inequalities obtained by this equation shows that the singular fiber is any of the types I_b $(b \geq 2)$, III, IV, I_b^*, II*, III*, IV*.

Finally, we discuss the multiplicity. Let m be the multiplicity of the fiber $C(a)$. If $C(a)$ is simply-connected, then the multiplicity is $m = 1$ by the following argument. Let Δ be a small disk on S centered at a, and $f_a : \Delta \to \mathbb{C}$ a holomorphic coordinate function with $f_a(a) = 0$. Now we cover $C(a)$ by a finite number of small open sets N_i so that if $N_i \cap N_j \neq \emptyset$, then $N_i \cap N_j \cap C(a) \neq \emptyset$. Then, there is a holomorphic function h_i on each N_i such that $h_i^m = f_a \circ \Phi$. Since $h_i = c_{ij} h_j$ holds on $N_i \cap N_j$ for some constant c_{ij} with $c_{ij}^m = 1$, we obtain a collection $\{c_{ij}\}$ which satisfies the cocycle condition. Since we have $H^1(C(a), \mathbb{Z}_m) = 0$ by the simply-connectedness of $C(a)$, the 1-cocycle (c_{ij}) is coboundary, namely, there exists a collection of constants c_i with $c_i^m = 1$ such that $c_{ij} = c_j c_i^{-1}$. Therefore, by setting $h = c_i h_i$ on N_i, we obtain a holomorphic function h on $\Phi^{-1}(\Delta)$ satisfying $h^m = f_a \circ \Phi$. Then, $\Phi^{-1}(u)$ consists of m components for any $u \in \Delta$ with $u \neq a$, while it is a smooth elliptic curve since $\Phi^{-1}(u)$ is a regular fiber of Φ. Therefore, the multiplicity m must be 1.

The only fibers that are not simply-connected are I_0, I_1 and I_b. In these cases, the multiplicity m is arbitrary, and $C(a)$ may be a multiple fiber $_mI_0$, $_mI_1$ or $_mI_b$. In the other cases, $m = 1$ and a multiple fiber does not appear. □

For a singular fiber of an elliptic fibration, the *monodromy* is defined to be that of the T^2-bundle over S^1 obtained as the preimage of a small loop around the corresponding critical value. Kodaira calculated the monodromy around each singular fiber by giving the local model for each case ([97], Table I). Here we will explain only the case of an I_1 fiber, and after that, give the local model for an $_mI_0$ fiber in the explanation of logarithmic transformations.

First we describe the local model of an I_1 fiber in two ways. In what follows, we make use of the following notations:

$$\Delta(r) := \{z \in \mathbb{C} \mid |z| < r\}, \quad \Delta(r_1, r_2) := \{z \in \mathbb{C} \mid r_1 < |z| < r_2\}.$$

Moreover, ρ always denotes a real constant with $0 < \rho < 1$. First we recall the Weierstrass normal form of a projective cubic curve. An elliptic curve $\mathbb{C}/\mathbb{Z} \oplus \tau\mathbb{Z}$ with $\mathrm{Im}\,\tau > 0$ can be embedded in $\mathbb{C}P^2$ by the cubic equation

$$y^2 = 4x^3 - g_2 x - g_3,$$

where

$$g_2 = 60 \sum_{(m,n)\neq(0,0)} (m + n\tau)^{-4},$$

$$g_3 = 140 \sum_{(m,n)\neq(0,0)} (m + n\tau)^{-6}.$$

The embedding is given by the so-called *Weierstrass $\wp$-function* $\wp(z)$ over the elliptic curve. Since it satisfies the differential equation

$$\wp'(z)^2 = 4\wp(z)^3 - g_2\wp(z) - g_3,$$

the map $z \mapsto [x : y : 1] = [\wp(z) : \wp'(z) : 1]$ indeed gives an embedding of the elliptic curve. From this point of view, a cubic equation is said to be in *Weierstrass normal form* if it is in the form $y^2 = x^3 - ax - b$. The corresponding cubic curve is a smooth elliptic curve if and only if the discriminant $a^3 - 27b^2$ is non-zero. Based on this knowledge, it is obvious that the equation

$$y^2 = 4x^3 - p(t)x - q(t),$$

where $p(t)$ and $q(t)$ are holomorphic functions of $t \in \Delta(\rho)$, gives an elliptic fibration over the small disk $\Delta(\rho)$. For each t, the equation defines a possibly singular elliptic curve, which we call a *Weierstrass elliptic curve*. It is a singular elliptic curve if and only if the modular discriminant function $D(t) = p(t)^3 - 27q(t)^2$ vanishes, and has an ordinary double point if and only if $D(t)$ has a simple zero at the point. We call such a local elliptic fibration the *Weierstrass model*.

For the description of the local model of an I_1 fiber, Kodaira uses a local elliptic fibration in a slightly generalized form, which we still call a Weierstrass model. Let

$$S = \left\{([z_0 : z_1 : z_2], \tau) \mid z_1^2 z_2 - 4z_0^3 - z_0^2 z_2 + \alpha_2(\tau)z_0 z_2^2 + \alpha_3(\tau)z_2^3 = 0\right\}$$

$$\subset \mathbb{C}P^2 \times \Delta(\rho),$$

where

$$\alpha_2(\tau) = 20 \sum_{n=1}^{\infty} (1 - \tau^n)^{-1} n^3 \tau^n,$$

$$\alpha_3(\tau) = \frac{1}{3} \sum_{n=1}^{\infty} (1 - \tau^n)^{-1} (7n^5 + 5n^3) \tau^n.$$

Since the curve $\{z_1^2 z_2 - 4z_0^3 - z_0^2 z_2 + \alpha_2(\tau) z_0 z_2^2 + \alpha_3(\tau) z_2^3 = 0\}$ is a non-singular elliptic curve if $\tau \neq 0$ and has an ordinary double point at $[0 : 0 : 1]$ if $\tau = 0$, the canonical projection $p \colon \mathbb{C}P^2 \times \Delta(\rho) \to \Delta(\rho)$ restricts to a holomorphic map $p|_S \colon S \to \Delta(\rho)$, which is an elliptic fibration with one I_1 singular fiber $(p|_S)^{-1}(0)$. This is the local model for an I_1 fiber written by a Weierstrass model.

Next, we see another description using the quotient by a group action. This model is more suitable for the usage in the later application (see Chap. 5) in the sense that it is easy to describe all the holomorphic sections (Proposition 2.3) and to see the monodromy around the singular fiber. First we fix a non-zero complex number τ with $|\tau| \neq 1$, and take a free action of $\mathbb{Z}$ to $\mathbb{C}^*$ defined by $1 \cdot z = \tau z$. Then the quotient $\mathbb{C}^*/\mathbb{Z}$ is an elliptic curve with modulus $\frac{1}{2\pi i} \log \tau$, since the multi-valued holomorphic map $\frac{1}{2\pi i} \log \colon \mathbb{C}^* \to \mathbb{C}$ gives the lattice $\mathbb{Z} \oplus \frac{1}{2\pi i} \log \tau \mathbb{Z} \subset \mathbb{C}$. Now, for any $0 < \rho < 1$, we consider the elliptic fibration

$$\pi \colon \mathbb{C}^* \times \Delta(0, \rho)/\mathbb{Z} \to \Delta(0, \rho),$$

where the action of $\mathbb{Z}$ is given by

$$n \cdot (z, t) = (t^n z, t),$$

and π is the map induced by the second projection. Then the fiber over t is a non-singular elliptic curve of modulus $\frac{1}{2\pi i} \log t$. Hence, it is a non-singular elliptic fibration over $\Delta(\rho, 0)$. By setting $t = \varepsilon e^{2\pi i \theta}$ with $0 < \varepsilon < \rho$ and observing how the lattice $\mathbb{Z} \oplus \frac{1}{2\pi i} \log t \mathbb{Z}$ changes as θ varies from 0 to 1, it is easily proven that the monodromy is given by the matrix $\begin{pmatrix} 1 & 1 \\ 0 & 1 \end{pmatrix}$.

This fibration extends over $\Delta(\rho)$ as a singular elliptic fibration whose singularity is an ordinary double point. Indeed, Kodaira gave an explicit biholomorphism between $\mathbb{C}^* \times \Delta(0, \rho)/\mathbb{Z}$ and $(p|_S)^{-1}(\Delta(0, \rho))$ by putting

$$z_0/z_2 = -\frac{1}{12} - \frac{1}{4\pi^2} \wp\left(\frac{1}{2\pi i} \log z\right) \quad \text{and} \quad z_1/z_2 = \frac{i}{8\pi^3} \wp'\left(\frac{1}{2\pi i} \log z\right).$$

Hence, we obtain the completion W of $\mathbb{C}^* \times \Delta(0, \rho)/\mathbb{Z}$. Now we have a singular elliptic fibration $g \colon W \to \Delta(\rho)$, which is another description of the local model for an I_1 fiber.

Proposition 2.3 *Any holomorphic section of the elliptic fibration*

$$\pi : \mathbb{C}^* \times \Delta(0, \rho)/\mathbb{Z} \to \Delta(0, \rho)$$

can be described by the multi-valued holomorphic function

$$\varphi(t) = \exp \left(h(t) - \frac{m}{2} \log t + \frac{m}{4\pi i} (\log t)^2 \right),$$

where $m \in \mathbb{Z}$ and h is an arbitrary single-valued holomorphic function on $\Delta(0, \rho)$.

Proof Let σ be a holomorphic section of π, and φ be the corresponding multi-valued holomorphic function from $\Delta(0, \rho)$ to $\mathbb{C}^*$. Notice that the holomorphic function $\exp(2\pi i \cdot) \colon \mathbb{C} \to \mathbb{C}^*$ is a covering map and its restriction gives a covering map from $\left\{ s \in \mathbb{C} \mid \operatorname{Im} s < \frac{-1}{2\pi} \log \rho \right\}$ to $\Delta(0, \rho)$. Hence, there is a single-valued holomorphic function $f \colon \left\{ s \in \mathbb{C} \mid \operatorname{Im} s < \frac{-1}{2\pi} \log \rho \right\} \to \mathbb{C}$ satisfying $\exp 2\pi i f(s) = \varphi(e^{2\pi i s})$. Then we have

$$f(s + 1) = f(s) + ms + n$$

for some integers m and n. We define a holomorphic function g by

$$g(s) = ns + m \frac{s(s-1)}{2} = (n - \frac{m}{2})s + \frac{m}{2} s^2.$$

Then it satisfies $g(s+1) = g(s) + ms + n$, and hence, $f - g$ is a periodic function of period 1 defined on $\left\{ s \in \mathbb{C} \mid \operatorname{Im} s < \frac{-1}{2\pi} \log \rho \right\}$. Thus we can define a holomorphic function h on $\Delta(0, \rho)$ by $h(t) = 2\pi i (f(s) - g(s))$, where $t = e^{2\pi i s}$. Hence, we obtain

$$\varphi(t) = \exp 2\pi i f(s) = t^n \exp \left(h(t) - \frac{m}{2} \log t + \frac{m}{4\pi i} (\log t)^2 \right).$$

However, all φ's with the same m and h give the same section of π, since the factor t^n is absorbed by the $\mathbb{Z}$-action on $\mathbb{C}^* \times \Delta(0, \rho)$. Therefore, it is enough to consider the case where $n = 0$ and we have

$$\varphi(t) = \exp \left(h(t) - \frac{m}{2} \log t + \frac{m}{4\pi i} (\log t)^2 \right).$$

This completes the proof. $\square$

Finally, we review the relation between logarithmic transformations and multiple fibers of elliptic surfaces.

Let $\Phi \colon V \to C$ be an elliptic surface and $\Phi^{-1}(a)$ its regular fiber over a point $a \in C$. Then there exists a small neighborhood Δ of a such that Δ is biholomorphic to $\{t \in \mathbb{C} \mid |t| < 1\}$ and $\Phi^{-1}(\Delta)$ contains no singular fiber. The non-singular local elliptic fibration $\Phi|_{\Phi^{-1}(\Delta)} \colon \Phi^{-1}(\Delta) \to \Delta$ can be represented in the form

$$\pi \colon \mathbb{C} \times \Delta/(\mathbb{Z} \oplus \mathbb{Z}\omega(t)) \to \Delta,$$

where ω is a holomorphic function on Δ such that $\operatorname{Im}\omega(t) > 0$, and π is the map induced by the second projection. Now let $\psi \colon \Delta \to \Delta$ be the map given by $\psi(s) = s^m$ for a positive integer m. Taking the pullback of the elliptic fibration π by ψ, we obtain a new fibration

$$\mathbb{C} \times \Delta/(\mathbb{Z} \oplus \mathbb{Z}\omega(s^m)) \to \Delta; (z, s) \mapsto s,$$

whose total space we denote by X. Then X admits the free $\mathbb{Z}_m$-action generated by

$$(z, s) \mapsto \left(z + \frac{k}{m}, \ e^{\frac{2\pi i}{m}} s\right),$$

where k is an integer coprime with m. We denote by Y the quotient of X by this $\mathbb{Z}_m$-action. Then the map $\Psi \colon Y \to \Delta$ defined by $\Psi([(z, s)]) = s^m$ is an elliptic fibration such that the only singular fiber is $\Psi^{-1}(0)$, which is obviously a multiple fiber of type $_m I_0$. Notice that $\Psi^{-1}(\Delta - \{0\})$ is isomorphic to $\Phi^{-1}(\Delta - \{0\})$ as an elliptic fibration. Indeed, the map defined by

$$[(z, s)] \mapsto \left(z - \frac{k}{2\pi i} \log s, \ s^m\right)$$

gives the isomorphism. Hence, by replacing $\Phi^{-1}(\Delta)$ by $Y = \Psi^{-1}(\Delta)$, we obtain a new elliptic fibration over C with a multiple fiber on $a \in C$. This operation is called a *logarithmic transformation of multiplicity* m.

Remark 2.1 In the above definition, the logarithmic transformation of multiplicity 1 is nothing but the trivial operation. According to Kodaira's terminology, however, the following operation is also called a logarithmic transformation of multiplicity 1.

Let $E = \mathbb{C}/\mathbb{Z} \oplus \tau\mathbb{Z}$ for a complex number τ with $\operatorname{Im}\tau > 0$. Take the product $E \times \mathbb{C}P^1$. Then the second projection is the trivial elliptic bundle. This trivial elliptic surface can be described as the union of the two pieces $E \times \mathbb{C}_z$ and $E \times \mathbb{C}_w$ glued by the identification map $E \times (\mathbb{C}_z - \{0\}) \to E \times (\mathbb{C}_w - \{0\})$ given by $(\xi, z) \mapsto (\eta, w) = (\xi, \frac{1}{z})$. Now we change the gluing map to the map

$$(\xi, z) \mapsto (\eta, w) = \left(\xi - \frac{1}{2\pi i} \log z, \ \frac{1}{z}\right).$$

Then we obtain a new elliptic bundle over $\mathbb{C}P^1$, whose total space is diffeomorphic to $S^1 \times S^3$. In fact, this is the special case of a primary Hopf surface, which can be obtained by taking the quotient of $\mathbb{C}^2 - \{(0,0)\}$ by the $\mathbb{Z}$-action generated by $(z_1, z_2) \mapsto (e^{2\pi i \tau} z_1, e^{2\pi i \tau} z_2)$. For the detail of Hopf surfaces, see Sect. 2.3.

We close this section by mentioning the two important results.

Theorem 2.3 (Theorem 6.3 in [97]) *Let* $\Phi \colon V \to C$ *be an elliptic fibration with multiple fibers. Then there exists an elliptic surface* $\tilde{V}$ *without multiple fibers, which can be obtained by taking a branched covering of* V *whose branch curves are regular fibers of* Φ.

Theorem 2.4 (Corollary to Theorem 11.3 in [98]) *An elliptic surface without multiple fibers is deformation equivalent to an elliptic surface that admits a holomorphic section.*

2.3 Compact Non-Kähler Surfaces

For a compact Kähler manifold, the odd-dimensional Betti number must be even by the Hodge theory (Theorem 1.10). In the case of compact complex surfaces, the converse is also true.

Theorem 2.5 (Kodaira, Miyaoka, Siu) *A compact complex surface is Kähler if and only if its first Betti number is even.*

This theorem is conjectured by Kodaira, and proved by Miyaoka and Siu.

Theorem 2.6 (Miyaoka [130]) *An elliptic surface admits a Kähler metric if and only if its first Betti number is even.*

Theorem 2.7 (Siu [165]) *Every K3 surface admits a Kähler metric.*

By Theorem 2.5, a compact non-Kähler surface is nothing but a complex surface whose first Betti number is odd. According to Kodaira's terminology, they are of class VI or of class VII. Since surfaces of class VI are elliptic surfaces, here we will focus on the surfaces of class VII, especially, minimal ones. A minimal surface of class VII is said to be of *class VII$_0$*. There are several important examples in this class.

Hopf Surfaces
Definition 2.2 (Hopf Surface) A compact complex surface is called a *Hopf surface* if its universal covering space is $\mathbb{C}^2 - \{(0,0)\}$.

Hopf surfaces are the most fundamental examples of compact non-Kähler surfaces, and they are examined in detail by Kodaira [100].

Let V be a Hopf surface. Then it can be expressed as $V = \mathbb{C}^2 - \{(0,0)\}/G$, where the group G consists of some automorphisms of $\mathbb{C}^2 - \{(0,0)\}$. By Hartogs' lemma, an element of G extends over $\mathbb{C}^2$ as an automorphism fixing the origin. Since G contains a contraction that belongs to the center of G, we choose such an element g. Then the quotient of $\mathbb{C}^2 - \{(0, 0)\}$ by the normal subgroup of G generated by g forms a finite unramified covering of V. On the other hand, by a result of Lattès [106] and Sternberg [167], the contraction g can be expressed in the form

$$g(z_1, z_2) = (\alpha_1 z_1 + \beta z_2^m, \alpha_2 z_2),$$

where $0 < |\alpha_1| \le |\alpha_2| < 1$ and $(\alpha_1 - \alpha_2^m)\beta = 0$. From these viewpoints, Hopf surfaces are divided into two classes.

Definition 2.3

1. A *primary Hopf surface* is a complex surface defined as the quotient of the action of $\mathbb{Z}$ on $\mathbb{C}^2 - \{(0, 0)\}$ generated by g.
2. A *secondary Hopf surface* is a Hopf surface that is not primary.

Since the action of a contraction g gives the fundamental domain $B^4 - g(B^4)$ that is diffeomorphic to $S^3 \times (0, 1]$, a primary Hopf surface is diffeomorphic to $S^3 \times S^1$. In particular, its first Betti number is equal to 1. A secondary Hopf surface V also satisfies $b_1(V) = 1$, since there exists a primary Kodaira surface which is a finite unramified covering of V. Therefore any Hopf surface is a surface of class VII_0. In [100], Kodaira gave various characterizations of Hopf surfaces.

Theorem 2.8 (Theorem 31 in [100]) *A primary Hopf surface is an elliptic surface if and only if $\beta = 0$ and $\alpha_1^k = \alpha_2^l$ for some positive integers k and l.*

Indeed, if V is a primary Hopf surface with $\beta = 0$ and $\alpha_1^k = \alpha_2^l$ for some k and l, then we can define a surjective holomorphic map $f : V \to \mathbb{C}P^1$ by $f(z_1, z_2) = [z_1^k : z_2^l]$. It can be explicitly checked that f is an elliptic fibration. The converse is proven by an elementary (but a little bit more complicated) argument.

Theorem 2.9 *For a compact complex surface V, the following three conditions are equivalent:*

(1) V is a primary Hopf surface.
(2) $b_2(V) = 0$ and $\pi_1(V) = \mathbb{Z}$.
(3) V is diffeomorphic to $S^3 \times S^1$.

It is obvious that (1) implies (3), and hence (2). The difficult part is to show that (2) implies (1). Since a surface V with $b_2(V) = 0$ and $\pi_1(V) = \mathbb{Z}$ is of class VII_0, there are two possibilities $a(V) = 0$ and $a(V) = 1$. If $a(V) = 1$, then V is an elliptic surface of class VII_0, and hence, it is obtained from the product $\mathbb{C}P^1 \times C$ by a finite iteration of logarithmic transformations, where C is an elliptic curve (Theorem 27 in [100]). This classification of elliptic surfaces of class VII_0 leads to the conclusion. In the case $a(V) = 0$, the key of the proof is the following result.

Theorem 2.10 (Theorem 34 in [100]) *Let V be a compact complex surface with $a(V) = 0$, $b_1(V) = 1$ and $b_2(V) = 0$. If V contains at least one curve, then V is a Hopf surface.*

Finally, Kodaira proved that (2) implies the existence of a compact curve, which is the most difficult part, and completed the proof of Theorem 2.9.

Kodaira Surfaces
Definition 2.4 (Kodaira Surface)

1. Let L be a line bundle over an elliptic curve E with non-trivial Chern class and L^* its associated $\mathbb{C}^*$ bundle. Take the quotient of the action to L^* generated by multiplication by non-zero complex number τ with $|\tau| \neq 1$. Then we obtain an elliptic bundle over the elliptic curve E, since each fiber is the quotient $\mathbb{C}^*/\mathbb{Z}$, which is the elliptic curve of modulus $\frac{1}{2\pi i} \log \tau$. Thus we obtain a compact complex surface as the total space of the elliptic bundle. This is called a *primary Kodaira surface*.
2. A primary Kodaira surface sometimes admits a free action of a finite group of automorphisms. A compact complex surface obtained as the quotient by such an action is called a *secondary Kodaira surface*.

A primary Kodaira surface is of class VI and a secondary Kodaira surface is of class VII, since their first Betti numbers are 3 and 1, respectively. This is verified as follows.

First, let V be a primary Kodaira surface and L the corresponding line bundle over the elliptic curve E with $\deg(L) = n$. Then, the S^1-bundle associated with L is isomorphic to the S^1-bundle over the 2-torus T^2 with Euler class $n \in H^2(T^2; \mathbb{Z}) = \mathbb{Z}$. Denoting the total space of this bundle by N^3, it is easily checked that $\pi_1(N^3) = \mathbb{Z}^2 \oplus \mathbb{Z}_{|n|}$. Indeed, N^3 also fibers over S^1 with fiber T^2 and monodromy $\begin{pmatrix} 1 & n \\ 0 & 1 \end{pmatrix}$. (Let M_n be the mapping torus of $\begin{pmatrix} 1 & n \\ 0 & 1 \end{pmatrix}$, that is, $M_n = S^1 \times S^1 \times \mathbb{R}/(x + ny, y, t) \sim (x, y, t + 1)$, where (x, y) and t are coordinates on $S^1 \times S^1$ and $\mathbb{R}$, respectively. Then consider the map $M_n \to S^1 \times S^1; [(x, y, t)] \mapsto (y, [t])$.) Such a closed 3-manifold is called a *nil 3-manifold*. Thus the total space of L^* is diffeomorphic to the product of $\mathbb{R}$ and the nil 3-manifold N^3. Taking the quotient of L^* by the $\mathbb{Z}$-action, we obtain the primary Kodaira surface V, which is diffeomorphic to the product $S^1 \times N^3$. Therefore, it follows that $\pi_1(V) = \mathbb{Z}^3 \oplus \mathbb{Z}_{|n|}$, and hence, $b_1(V) = 3$.

Now let us consider which free finite actions a primary Kodaira surface admits. The only possible actions are cyclic ones of order 2, 3, 4 or 6, since the automorphisms of the base elliptic curve E up to translations form $\mathbb{Z}_4$, $\mathbb{Z}_6$ or $\mathbb{Z}_2$ if the corresponding lattice is $\mathbb{Z} \oplus \mathbb{Z}i$, $\mathbb{Z} \oplus \mathbb{Z}e^{\frac{\pi i}{3}}$ or otherwise, respectively. For each case, there exists a primary Kodaira surface V such that the cyclic group acts freely on V as automorphisms. Here we explain only the case of order 2. For simplicity, we assume that $E = \mathbb{C}/\mathbb{Z} \oplus \mathbb{Z}i$. Then the multiplication by -1 defines an involution ι of E with just 4 fixed points, which correspond to the lattice $\frac{1}{2}\mathbb{Z} \oplus \frac{1}{2}\mathbb{Z}i$. Now, denoting by p the point on E corresponding to the origin, we take the line bundle $L = [np]$ and the corresponding primary Kodaira surface V. Then there exists an automorphism g of the total space of L that covers $\iota: E \to E$. On the other hand, the multiplication on fiber complex lines by $\tau^{\frac{1}{2}}$ defines an automorphism of V that covers the identity map id_E. Composing it with g, we obtain a fixed point free involution of V covering ι. Then, the quotient by the $\mathbb{Z}_2$-action generated by this involution is a secondary Kodaira surface. By construction, it admits an elliptic fibration over $\mathbb{C}P^1$ with just 4 multiple fibers of type $_2I_0$. Now it is easy to check that the first Betti number of this elliptic surface is equal to 1.

Remark 2.2 From the viewpoint of differential topology, a primary Kodaira surface is called a *Kodaira-Thurston manifold*. This is diffeomorphic to the product of S^1 and a nil 3-manifold (as we have seen above), and is famous as an example that admits both complex structures and symplectic structures, but no Kähler ones.

Inoue Surfaces

Inoue [77] constructed compact complex surfaces S such that $b_1(S) = 1, b_2(S) = 0$ and S has no compact curves. Such surfaces (of class VII$_0$) are called *Inoue surfaces*. Here we review his construction.

Let $M = (m_{ij})$ be an element in $GL(3; \mathbb{Z})$ with $\det M = 1$. Suppose M has one real eigenvalue α and two imaginary eigenvalues $\beta, \bar{\beta}$, where $\bar{\beta}$ denotes the complex conjugate of β. We assume further that $\alpha > 0, \alpha \neq 1$ and $\mathrm{Im}\beta > 0$. Take eigenvectors (a_1, a_2, a_3) and (b_1, b_2, b_3) corresponding to α and β, respectively, so that (a_1, a_2, a_3) is a real vector. We note that $(\bar{b}_1, \bar{b}_2, \bar{b}_3)$ is an eigenvector belonging to $\bar{\beta}$.

Let (z, w) be the coordinates on $\mathbb{H} \times \mathbb{C}$, where $\mathbb{H}$ denotes the upper half plane. Now we define analytic automorphisms g_0, g_1, g_2, g_3 of $\mathbb{H} \times \mathbb{C}$ as follows:

$$g_0(z, w) = (\alpha z, \beta w)$$

$$g_i(z, w) = (z + a_i, w + b_i) \quad (i = 1, 2, 3).$$

Then the automorphism group G_M^+ generated by g_0, g_1, g_2, g_3 acts freely and properly discontinuously on $\mathbb{H} \times \mathbb{C}$. Hence the quotient $S_M^+ = \mathbb{H} \times \mathbb{C}/G_M^+$ is a compact complex surface. If we adopt $\bar{\beta}$ and $(\bar{b}_1, \bar{b}_2, \bar{b}_3)$ instead of β and (b_1, b_2, b_3), we obtain another automorphism group G_M^- and a complex surface $S_M^- = \mathbb{H} \times \mathbb{C}/G_M^-$. Then S_M^+ and S_M^- are not biholomorphic to each other. However, they are diffeomorphic, since the diffeomorphism of $\mathbb{H} \times \mathbb{C}$ given by $(z, w) \mapsto (z, \bar{w})$ induces a diffeomorphism $S_M^+ \to S_M^-$. Let us denote the underlying manifold by X_M. Since the automorphism groups G_M^+ and G_M^- have the representation

$$\left\langle g_0, g_1, g_2, g_3 \mid g_i g_j = g_j g_i, \ g_0 g_i g_0^{-1} = g_1^{m_{i1}} g_2^{m_{i2}} g_3^{m_{i3}}, \ 1 \leq i, j \leq 3 \right\rangle,$$

X_M is diffeomorphic to the T^3-bundle over S^1 with monodromy M. In particular, $b_1(X_M) = 1$ and $b_2(X_M) = 0$. The non-existence of compact curves follows from the next proposition. Let $\Gamma^\pm$ be the normal subgroup of $G_M^\pm$ generated by g_1, g_2, g_3.

Proposition 2.4 (Kodaira, see also Lemma 3 in [77]) *Any* Γ^+ *(resp.* Γ^-*) invariant holomorphic function on* $\mathbb{H} \times \mathbb{C}$ *must be constant.*

For further details, see [77–79].

Inoue–Hirzebruch Surfaces

A normal complex surface singularity is called a *cusp singularity* if the exceptional set of its minimal resolution is a cycle of rational curves. In [72], Hirzebruch explicitly constructed the minimal resolution of a cusp singularity. Based on this result, Inoue [80] constructed compact complex surfaces of VII_0 with two cycles of rational curves, which are now called *Inoue–Hirzebruch surfaces* or *hyperbolic Inoue surfaces*.

Let K be a totally real quadratic field over $\mathbb{Q}$ and M an additive subgroup of K which is a free abelian group of rank 2. Let U_M^+ be the subgroup of K defined by

$$U_M^+ = \{x \in K \mid x > 0, \ \bar{x} > 0, \ xM = M\}.$$

Here $\bar{x}$ denotes the conjugate irrational number of x (recall that x is a quadratic irrational number). Then U_M^+ is an infinite cyclic group. Now we take a non-trivial subgroup V of U_M^+ and define

$$G(M, V) = \left\{ \begin{pmatrix} e & a \\ 0 & 1 \end{pmatrix} \middle| e \in V, a \in M \right\}.$$

Then $G(M, V)$ acts on $\mathbb{H} \times \mathbb{C}$ by

$$\begin{pmatrix} e & a \\ 0 & 1 \end{pmatrix} \cdot (z_1, z_2) = (e z_1 + a, \bar{e} z_2 + \bar{a}).$$

This action is free and properly discontinuous, and hence, the quotient $\mathbb{H} \times \mathbb{C}/G(M, V)$ is a complex surface. This manifold is not compact, but it can be compactified to a normal complex space $Y(M, V)$ by adding two points ∞ and ∞_- and giving an appropriate complete system of neighborhoods of each point. In fact, the two points are cusp singularities and their minimal resolutions are explicitly given in [72]. Hence, we can take the minimal resolution $\pi : S(M, V) \to Y(M, V)$ and obtain a non-singular compact complex surface $S(M, V)$. This is called an *Inoue–Hirzebruch surface*. The complex surface $S(M, V)$ contains two cycles of rational curves $C = C_1 + \cdots + C_k$ and $D = D_1 + \cdots + D_l$ as the exceptional sets of the minimal resolution of ∞ and ∞_-, respectively.

Nakamura [134] showed the duality between the two cycles C and D in terms of the repeating modified continued fractions obtained from the self-intersection numbers of the rational curves C_i and D_j. This interesting duality can be seen as a variant of the strange duality between the 14 exceptional unimodal singularities found by Arnol'd. Nakamura also proved that any smooth deformation of a singular Inoue–Hirzebruch surface $Y(M, V)$ is a $K3$ surface. In this direction, Kodama, Mitsumatsu, Mori and the author [89] recently proved that for each of 10 pairs of hypersurface cusp singularities appearing in Nakamura's duality pairs, the corresponding two Milnor fibers can be smoothly glued together along the boundaries to become a $K3$ surface.

Other Surfaces

Inoue constructed further examples of surfaces of class VII_0 called *half Inoue surfaces* and *parabolic Inoue surfaces*. A half Inoue surface is nothing but the quotient of some Inoue–Hirzebruch surface by a fixed point free involution. On the other hand, parabolic Inoue surfaces were later generalized by Enoki [40] as so-called *Enoki surfaces*. There are also examples by Kato [90, 91], called *Kato surfaces*. A *Kato surface* is a surface of class VII_0 with positive b_2 containing a so-called *global spherical shell*, which is defined as a domain that is biholomorphic to the standard neighborhood of S^3 in $\mathbb{C}^2$ and does not separate the surface. For the details of these examples, see [135, 136].

The classification of surfaces of class VII_0 is an important, but very difficult problem. Towards the solution to it, Nakamura [136] proposed the conjecture that any class VII_0 surface with $b_2 > 0$ is a Kato surface. Indeed, every known example of such a surface is a Kato surface [31, 90, 91]. Though the conjecture itself is considered by experts to be difficult to access, Teleman (and others) made some progress in this direction by a gauge theoretical approach. They classified surfaces of class VII_0 with $b_2 \leq 3$ in the following sense.

1. A surface of class VII_0 with $b_2 = 0$ is biholomorphic to either a Hopf surface or an Inoue surface ([111, 170], see also [11, 12]).
2. A surface of class VII_0 contains an elliptic curve or a cycle of rational curves if $b_2 = 1, 2$ or 3. Moreover, such surfaces are classified up to deformation equivalences or up to diffeomorphisms [171, 172].

For more details, see [173].

Chapter 3
Elliptic Surfaces and Lefschetz Fibrations

In the last chapter, we reviewed the classification of compact complex surfaces, especially Kodaira's work about elliptic surfaces and the classification of singular fibers. In this chapter, we focus on the counterpart of elliptic surfaces in differential topology, that is, genus-one Lefschetz fibrations. It played an important role in the work of Kas and Moishezon which determined the diffeomorphism types of elliptic surfaces over $\mathbb{C}P^1$, and was completed by the theorem of Matsumoto.

3.1 Definition of Lefschetz Fibrations

First, let us recall the definition of Lefschetz fibrations on 4-manifolds.

Definition 3.1 (Lefschetz Fibration) Let M and B be closed connected oriented manifolds of dimension 4 and 2, respectively. A C^∞ smooth map $f : M \to B$ is called a *Lefschetz fibration* if it satisfies the following three conditions:

(1) f has a finite number of critical values $q_1, \ldots, q_n$ and its restriction to $f^{-1}(B - \{q_1, \ldots, q_n\})$ is a smooth fiber bundle.
(2) Each singular fiber $f^{-1}(q_i)$ contains only one critical point.
(3) For each critical point p, there exist local complex coordinates (z_1, z_2), which respects the orientation of M, and w around $p \in M$ and $f(p) \in B$, respectively, such that f is locally expressed as $w = f(z_1, z_2) = z_1 z_2$.

A regular fiber $f^{-1}(q)$, the preimage of a regular value q, is a closed oriented surface Σ_g, whose genus g is called the *genus of the Lefschetz fibration* f. A critical point satisfying condition (3) is called a *Lefschetz singularity*. Moreover, we usually make the following assumption.

© The Editor(s) (if applicable) and The Author(s), under exclusive license
to Springer Nature Singapore Pte Ltd. 2025
N. Kasuya, *Non-Kähler Complex Surfaces and Strongly Pseudoconcave Surfaces*,
SpringerBriefs in Mathematics, https://doi.org/10.1007/978-981-96-3002-8_3

(4) no fiber of f contains a 2-sphere that is embedded in M with self-intersection -1.

This condition is called the *relative minimality*.

For a Lefschetz singularity $\mathbf{0}$ of $f(z_1, z_2) = z_1 z_2$, a nearby regular fiber is given by $z_1 z_2 = t$. Assuming $t > 0$, the fiber contains a simple loop $(\sqrt{t} e^{i\theta}, \sqrt{t} e^{-i\theta})$ with $\theta \in S^1$, which contracts to the critical point $\mathbf{0}$ as t goes to 0. Such a loop is called a *vanishing cycle* of the Lefschetz singularity. Taking the preimage $f^{-1}(S^1_\varepsilon)$ of a small circle S^1_ε around the critical value $0 \in \mathbb{C}$, we obtain a fiber bundle over S^1. The monodromy of this bundle is called the *monodromy* around the critical value. As is easily verified by the local model, it is the right-handed Dehn twist about the vanishing cycle.

Even in the general case, we can take a vanishing cycle c for any Lefschetz singularity p in a similar way. Indeed, taking a smooth path $l \colon [0, 1] \to B$ from the critical value $f(p)$ to a fixed regular value q so that l does not pass any other critical value of f, there exists a smooth family of loops c_t $(0 < t \le 1)$, each of which is a simple closed curve in the regular fiber $f^{-1}(l(t))$ and converges to the critical point p when t goes to 0. Then the loop $c = c_1$ on the regular fiber $f^{-1}(q) \cong \Sigma_g$ is a vanishing cycle of p, and the monodromy around $f(p)$ is the right-handed Dehn twist about c. Notice that when $g = 1$, this monodromy can be expressed by the matrix $\begin{pmatrix} 1 & 1 \\ 0 & 1 \end{pmatrix} \in SL(2; \mathbb{Z})$. It is not by chance that this coincides with the monodromy of an I_1 singular elliptic fiber in Sect. 2.2.

Indeed, if the map f is holomorphic, then a Lefschetz singularity is just an ordinary double point. In the case of elliptic fibrations, a (genus-one) Lefschetz singular fiber is exactly an I_1 singular fiber. Therefore, an elliptic fibration whose singular fibers are all of type I_1 is topologically nothing but a genus-one Lefschetz fibration. In fact, the most fundamental example of genus-one Lefschetz fibration over S^2 arises from algebraic geometry in the following manner.

Example 3.1 Let $z = [z_0 \colon z_1 \colon z_2]$ be the homogeneous coordinates on $\mathbb{C}P^2$. Take two generic cubic homogeneous polynomials $h_0(z_0, z_1, z_2)$ and $h_1(z_0, z_1, z_2)$. Then, by Bezout's theorem, the two elliptic curves $h_0^{-1}(0)$ and $h_1^{-1}(0)$ transversely intersect each other at nine points $p_1, \ldots, p_9$. Hence, for any $[t_0 \colon t_1] \in \mathbb{C}P^1$, the homogeneous cubic $t_0 h_0 + t_1 h_1$ defines an elliptic curve passing the nine points. Moreover, for any point $z \in \mathbb{C}P^2$ other than the nine points, there is only one elliptic curve which the point sits on. Indeed, the map

$$f \colon \mathbb{C}P^2 - \{p_1, \ldots, p_9\} \to \mathbb{C}P^1; \quad z \mapsto [-h_1(z) \colon h_0(z)]$$

gives the family of elliptic curves $\{t_0 h_0(z) + t_1 h_1(z) = 0\}$, $[t_0 \colon t_1] \in \mathbb{C}P^1$ that foliates $\mathbb{C}P^2$ except the nine points, which all the members pass. (Such f is called a *pencil* and $p_1, \ldots, p_9$ are called its *base points*.) By blowing up at the base points $p_1, \ldots, p_9$ of f, we obtain the holomorphic elliptic fibration $\Phi_1 \colon \mathbb{C}P^2 \# 9\overline{\mathbb{C}P^2} \to \mathbb{C}P^1$. Notice that this elliptic surface $\mathbb{C}P^2 \# 9\overline{\mathbb{C}P^2}$ is also rational. So, it is called a *rational elliptic surface* and is usually denoted by $E(1)$. By the genericity of h_0 and h_1, we may assume that all the singular fibers are of type I_1, that is, all the critical points are of Lefschetz type. Thus we obtain an example of genus-one Lefschetz fibration $\Phi_1 \colon E(1) \to S^2$.

The topological Euler characteristics of an I_1 singular fiber and $E(1) = \mathbb{C}P^2 \# 9 \overline{\mathbb{C}P^2}$ are 1 and 12, respectively. Hence, it follows that the Lefschetz fibration Φ_1 has just 12 singular fibers. The rational elliptic surface $E(1)$ plays a role as the unit of genus-one Lefschetz fibrations on S^2 with respect to the fiber-connected sum. Here the fiber-connected sum is the following operation to two Lefschetz fibrations of the same genus g, which produces a new genus g Lefschetz fibration.

Definition 3.2 (Fiber Connected Sum) Let $\pi_1 \colon M_1 \to B_1$ and $\pi_2 \colon M_2 \to B_2$ be genus g Lefschetz fibrations. For each j ($j = 1, 2$), take a regular value p_j of f_j and a small disk $D_j \subset B_j$ containing p_j. Then the restriction $\pi_j|_{\pi_j^{-1}(D_j)}$ is the trivial Σ_g bundle over D_j, and hence, each $\pi_j^{-1}(D_j)$ is diffeomorphic to the product $D_j \times \Sigma_g$. Now we set $B_j' = B_j - D_j$ and $M_j' = \pi_j^{-1}(B_j')$. Take an orientation-reversing diffeomorphism $\varphi \colon S^1 \cong \partial D_1 \to \partial D_2 \cong S^1$ and a fiber-preserving, orientation-reversing diffeomorphism $\Phi \colon \partial M_1' = \pi_1^{-1}(\partial D_1) \to \pi_2^{-1}(\partial D_2) = \partial M_2'$ covering φ. Gluing M_1' and M_2' by Φ, we obtain a new 4-manifold, which we denote by $M_1 \#_f M_2$. By construction, it admits a genus g Lefschetz fibration

$$\pi \colon M_1 \#_f M_2 \to B_1 \# B_2$$

whose restriction to M_j' coincides with $\pi_j|_{M_j'}$. This Lefschetz fibration π is called the *fiber-connected sum* of π_1 and π_2.

In general, the diffeomorphism type of $M_1 \#_f M_2$ depends on the choice of the diffeomorphism Φ. In the case $g = 1$, however, it does not matter if either π_1 or π_2 has two singular fibers whose monodromies are $\begin{pmatrix} 1 & 1 \\ 0 & 1 \end{pmatrix}$ and $\begin{pmatrix} 1 & 0 \\ -1 & 1 \end{pmatrix}$, respectively (Lemma 7 in [131], see also Lemma 8.3.6 in [61]). Since we only deal with cases of this sort, we don't let Φ appear in the notation of the fiber-connected sum. Also we have to pay attention to the point that this is merely a differential topological operation. Namely, even if π_1 and π_2 are holomorphic Lefschetz fibrations, $M_1 \#_f M_2$ is not necessarily a complex surface and π is only a smooth Lefschetz fibration from a smooth 4-manifold to a smooth surface.

Applying this operation to Example 3.1, we can make new genus-one Lefschetz fibrations over S^2. Namely, by defining inductively

$$E(2) = E(1) \#_f E(1), \quad \cdots, \quad E(m) = E(m-1) \#_f E(1),$$

we obtain a genus-one Lefschetz fibration $\Phi_m \colon E(m) \to S^2$ for any positive integer m, which has just $12m$ singular fibers.

In fact, Moishezon and Kas proved that there is no other genus-one Lefschetz fibration on S^2 (Theorem 3.2), which became a breakthrough point to the topological classification of elliptic surfaces. Now we are going to review the series of works by Kas, Moishezon and Matsumoto.

3.2 The Theorem of Kas and Moishezon

Moishezon [131] and Kas [84] independently classified the diffeomorphism types of elliptic surfaces over $\mathbb{C}P^1$ without multiple fibers. What they showed first is that a singular elliptic fiber can be split into singular fibers of type I_1 through a deformation of elliptic fibrations.

Theorem 3.1 Let $\Phi\colon V \to C$ be an elliptic fibration without multiple fibers. Then there exists a deformtaion of V to an elliptic surface $\widetilde{V}$ that has only I_1 singular fibers.

Outline of the Proof Since any elliptic surface with a singular fiber but without a multiple fiber is deformation equivalent to the one that has a holomorphic section by Theorem 2.4, we may assume that V originally has a holomorphic section $\sigma\colon C \to V$. First we contract all the components of reducible fibers that don't intersect the section σ. Then we obtain an elliptic surface V' with singularities and an elliptic fibration $\Phi'\colon V' \to C$ whose fibers are all irreducible. The singularities caused by the contraction are rational double points (simple singularities), since the rational curves of any reducible singular fiber forms the configuration of any of A_n, D_n, E_6, E_7, or E_8 if one component is excluded (see Theorem 2.2). By Kas' result [83], such an elliptic fibration can be expressed by the Weierstrass model, as a family of elliptic curves over C (see also [128]).

In order to construct such a deformation, it is enough to argue locally on a neighborhood of a singular fiber. Indeed, for a singular value x_0 of Φ', a local deformation constructed on a small disk Δ containing x_0 can be glued together with the trivial deformation defined outside a smaller disk Δ' with $x_0 \in \Delta'$ to obtain a deformation of the original compact elliptic surface V' (see pp. 118–121 of [131], see also Theorem 4.8 in [51]).

Since $\Phi'\colon V' \to C$ can be described by the Weierstrass model, it can be expressed over Δ as

$$\{([z_0 : z_1 : z_2], t) \in \mathbb{C}P^2 \times \Delta \mid z_0 z_2^2 - z_1^3 - p(t)z_0^2 z_1 - q(t)z_0^3 = 0\},$$

where $p(t)$ and $q(t)$ are holomorphic functions on Δ, and the curve

$$\left\{[z_0 : z_1 : z_2] \in \mathbb{C}P^2 \mid z_0 z_2^2 - z_1^3 - p(0)z_0^2 z_1 - q(0)z_0^3 = 0\right\}$$

is the only singular fiber. Let $a \in \mathbb{C}$ be a non-zero constant such that $ap(0) - q(0) \neq -a^3$, $ap(t) - q(t) \not\equiv 0$ and

$$p'(t)^3 q(t) + 2q'(t)^3 - ap'(t)\left(p'(t)^2 p(t) + 3q'(t)^2\right)$$

is a non-vanishing holomorphic function on Δ. Such a complex number a surely exists, since the disk Δ is small. Now we take linear deformations $p(t) + u$ and $q(t) + au$ of the holomorphic functions p and q with the deformation parameter u. Then the deformed elliptic fibration

$$z_0 z_2^2 - z_1^3 - (p(t) + u)z_0^2 z_1 - (q(t) + au)z_0^3 = 0$$

has only I_1 singular fibers for $u \neq 0$, if $|u|$ is small enough. Indeed, for such a parameter u, the discriminant function

$$D(t) = 4(p(t) + u)^3 + 27(q(t) + au)^2$$

has only simple zeros. This is verified as follows. Let t_0 be a zero of the function $D(t)$. Notice that the order of the zero is greater than 1 if and only if the two complex plane curves $c(t) = (p(t) + u, q(t) + au)$ and $4x^3 + 27y^2 = 0$ are tangential at the point $(p(t_0) + u, q(t_0) + au)$. Since the curve $4x^3 + 27y^2 = 0$ is parametrized by $x = -3s^2$, $y = -2s^3$, the tangential condition is described as

$$p(t) + u = -3s^2, \quad q(t) + au = -2s^3, \quad 6s(q'(t) - sp'(t)) = 0.$$

Suppose $s \neq 0$. Then, we have $q'(t) = sp'(t)$ from the third equation. Substituting it in the first and second equations, we obtain

$$p'(t)^2 p(t) + 3q'(t)^2 = -p'(t)^2 u, \quad p'(t)^3 q(t) + 2q'(t)^3 = -ap'(t)^3 u.$$

Hence, the tangential condition holds only if

$$p'(t)^3 q(t) + 2q'(t)^3 - ap'(t)\left(p'(t)^2 p(t) + 3q'(t)^2\right) = 0,$$

which does not happen by the assumption on the constant a. Next, we deny the possibility of the case $s = 0$. In this case, since the curve $c(t)$ is "tangential" to $4x^3 + 27y^2 = 0$ at the cusp $(0, 0)$, the following equations must hold:

$$p(t_0) + u = 0, \quad q(t_0) + au = 0, \quad q'(t_0) = 0.$$

Then we obtain

$$p'(t_0)^3 q(t_0) + 2q'(t_0)^3 - ap'(t_0)\left(p'(t_0)^2 p(t_0) + 3q'(t_0)^2\right)$$

$$= p'(t_0)^3 (q(t_0) - ap(t_0)) = 0,$$

which contradicts the assumption on a. Therefore, $D(t)$ has only simple zeros on Δ, and hence, all the singular fibers are of type I_1. Finally, taking the simultaneous resolution of singularities for the parameter u yields a desired deformation of the original local elliptic fibration $\Phi|_{\Phi^{-1}(\Delta)} \colon \Phi^{-1}(\Delta) \to \Delta$. $\qquad\qquad\square$

In other words, an elliptic surface without multiple fibers can be deformed into an elliptic Lefschetz fibration. Therefore, if one wants to know about the diffeomorphism type of an elliptic surface, it is enough to argue about that of the total space of genus-one Lefschetz fibration. Now we focus on the case where the base space is the 2-sphere.

Let $f \colon M \to S^2$ be a genus-one Lefschetz fibration and $q_1, \ldots, q_n$ its critical points. Then a loop γ_i on the base around each critical value $f(q_i)$ determines a T^2-bundle over S^1. Its monodromy is an automorphism of a regular fiber T^2, so it is expressed as an element in $SL(2; \mathbb{Z})$. Hence, we obtain the monodromy homomorphism

$$h \colon \pi_1(S^2 - \{q_1, \ldots, q_n\}) \to SL(2; \mathbb{Z}).$$

Obviously, it must satisfy the relation $h(\gamma_1) \cdots h(\gamma_n) = E_2$. Now we need to classify the homomorphisms with this relation.

Since $SL(2; \mathbb{Z})$ is generated by $A = \begin{pmatrix} 1 & 1 \\ 0 & 1 \end{pmatrix}$ and $B = \begin{pmatrix} 1 & 0 \\ -1 & 1 \end{pmatrix}$, the monodromy can be also expressed by only these two elements. Moreover, the group structure of $PSL(2; \mathbb{Z}) = SL(2; \mathbb{Z})/\{\pm E_2\}$ is the free product $(\mathbb{Z}/2\mathbb{Z}) * (\mathbb{Z}/3\mathbb{Z})$ generated by $BA^2 = \begin{pmatrix} 1 & 2 \\ -1 & -1 \end{pmatrix}$ of order 2, and $BA = \begin{pmatrix} 1 & 1 \\ -1 & 0 \end{pmatrix}$ of order 3. Thus the problem of describing the monodromy of genus-one Lefschetz fibrations can be reduced to some word problems.

Definition 3.3 (Elementary Transformation) Let G be a group, and G^n be the set of n-tuples $(g_1, \ldots, g_n)$ in G. The jth transformation is a map $R_j \colon G^n \to G^n$ defined by

$$R_j(g_1, \ldots, g_{j-1}, g_j, g_{j+1}, g_{j+2}, \ldots, g_n)$$

$$= (g_1, \ldots, g_{j-1}, g_{j+1}, g_{j+1}^{-1} g_j g_{j+1}, g_{j+2}, \ldots, g_n).$$

We call R_j and R_j^{-1} $(1 \le j \le n)$ the *elementary transformations*.

Notice that the braid group B_n acts on the set G^n of n-tuples in G by the so-called Hurewitz action. An elementary transformation R_j can be seen as a Hurewitz move corresponding to the generator σ_j of B_n.

Lemma 3.1 (Livne [131]) *Let* $G =< a, b \mid a^3 = b^2 = 1 >$ *and set* $s_0 = a^2 b$, $s_1 = aba$, $s_2 = ba^2$. *Suppose that* $g_1, g_2, \ldots, g_n$ *are conjugates of* $s_1 = aba$ *such that* $g_1 g_2 \cdots g_n = 1$. *Then, by a finite sequence of elementary transformations, the n-tuple* $(g_1, \ldots, g_n)$ *can be transformed to an n-tuple* $(h_1, \ldots, h_n)$ *with each* h_i *equal to* s_0, s_1 *or* s_2.

Using Livne's theorem, Moishezon proved the following.

Lemma 3.2 (Moishezon [131]) *Let* $g_1, g_2, \ldots, g_n \in SL(2; \mathbb{Z})$ *be conjugates of* $\begin{pmatrix} 1 & 1 \\ 0 & 1 \end{pmatrix}$ *such that* $g_1 g_2 \cdots g_n = \begin{pmatrix} 1 & 0 \\ 0 & 1 \end{pmatrix}$. *Then, n is even and the n-tuple* $(g_1, \ldots, g_n)$ *can be transformed by a finite sequence of elementary transformations to an n-tuple* $(h_1, \ldots, h_n)$ *with* $h_{2j-1} = \begin{pmatrix} 1 & 0 \\ -1 & 1 \end{pmatrix}$, $h_{2j} = \begin{pmatrix} 1 & 1 \\ 0 & 1 \end{pmatrix}$ $(1 \le j \le n/2)$.

Since $BA = \begin{pmatrix} 1 & 1 \\ -1 & 0 \end{pmatrix}$ has order 6 in $SL(2; \mathbb{Z})$, the even number n in the above theorem can be divided by 12. Hence, it follows that the number of singular fibers of a genus-one Lefschetz fibration over S^2 is a multiple of 12. Moreover, Lemma 3.2 implies that once the number $n = 12m$ is fixed, the isomorphism class of a Lefschetz fibration is uniquely determined. On the other hand, for any positive integer m, there exists a genus-one Lefschetz fibration $\Phi_m \colon E(m) \to S^2$, which has just $12m$ singular fibers. As a consequence, the following classification result is proven.

Theorem 3.2 (Kas [84], Moishezon [131]) *Let* $f \colon M \to S^2$ *be a relatively minimal genus-one Lefschetz fibration with at least one singular fiber. Then it is isomorphic to the Lefschetz fibration* $\Phi_m \colon E(m) \to S^2$, *where* $m = \chi(M)/12$.

Corollary 3.1 *The diffeomorphism type of a relatively minimal elliptic surface over* $\mathbb{C}P^1$ *without multiple fibers depends only on its Euler number.*

3.3 The Theorem of Matsumoto

Matsumoto generalized the threorem of Kas and Moishezon to the case where the base space is a surface of arbitrary genus.

Theorem 3.3 (Matsumoto [120]) *Let S be a closed oriented surface and* $f \colon M \to S$ *be a relatively minimal genus-one Lefschetz fibration with at least one singular fiber. Then it is isomorphic to the fiber-connected sum of the Lefschetz fibration* $\Phi_m \colon E(m) \to S^2$ *with the trivial torus bundle* $pr_1 \colon S \times T^2 \to S$, *where* $m = \chi(M)/12$ *and* pr_1 *is the projection to the first factor.*

The following is an immediate corollary of the theorem.

Corollary 3.2 *The diffeomorphism type of a relatively minimal elliptic surface without multiple fibers depends only on the genus of the base space and the Euler number of the total space.*

The outline of the proof of Theorem 3.3 is as follows. Let S be a closed oriented surface of genus g and $q_1, \ldots, q_n$ be points on S. Regarding S as the boundary of an embedded 3-dimensional handlebody in $\mathbb{R}^3$, the meridians are determined as homology classes and the longitudes can be chosen corresponding to them. Fix the base point $p \in S$, then the fundamental group of S is expressed as

$$\pi_1(S, p) = \langle l_1, \ldots, l_g, m_1, \ldots, m_g \mid [l_1, m_1] \cdots [l_g, m_g] = 1 \rangle,$$

where m_j and l_j are loops corresponding to the meridian and the longitude. Moreover, for each $i \in \{1, \ldots, n\}$, we take a loop γ_i around q_i with the base point p. Matsumoto refined the results of Livne and Moishezon so that the argument works even without the assumption $g_1 g_2 \cdots g_n = 1$, and he showed the following result, which proves Theorem 3.3 based on Lemma 3.2.

Lemma 3.3 *Let* $\rho \colon \pi_1(S \setminus \{q_1, \ldots, q_n\}, p) \to SL(2; \mathbb{Z})$ *be a homomorphism such that* $\rho(\gamma_1), \ldots, \rho(\gamma_n) \in SL(2; \mathbb{Z})$ *are conjugates of* $\begin{pmatrix} 1 & 1 \\ 0 & 1 \end{pmatrix}$. *Then, there exists an orientation-preserving homeomorphism* $h \colon (S \setminus \{q_1, \ldots, q_n\}, p) \to (S \setminus \{q_1, \ldots, q_n\}, p)$ *such that for any* $1 \le j \le g$,

$$\rho\left(h(l_j)\right) = \rho\left(h(m_j)\right) = \begin{pmatrix} 1 & 0 \\ 0 & 1 \end{pmatrix}.$$

A self-homeomorphism of S fixing the n points $q_1, \ldots, q_n$ is seen as an element of $\mathrm{Homeo}(S \setminus \{q_1, \ldots, q_n\})$ in a natural way. In particular, for each $i \in \{1, \ldots, n\}$, a homeomorphism that rotates q_i along a simple closed curve on S is an element of $\mathrm{Homeo}(S \setminus \{q_1, \ldots, q_n\})$. With the help of the assumption that the monodromy around each q_i is conjugate to $\begin{pmatrix} 1 & 1 \\ 0 & 1 \end{pmatrix}$, such a homeomorphism plays an important role in simplifying the appearance of the homomorphism ρ.

The heart of Theorem 3.3 (or Lemma 3.3) becomes clearer from the following observation. Let $f \colon X \to \Sigma_g$ be any orientable T^2-bundle over a closed oriented surface of genus g. Taking the fiber-connected sum of f and Φ_m, we obtain a new genus-one Lefschetz fibration $h \colon X \#_f E(m) \to \Sigma_g$. By Theorem 3.3, however, h is isomorphic to the connected sum of the trivial T^2-bundle over Σ_g and Φ_m. This implies that no matter how twisted the T^2-bundle f is (it may have non-trivial monodromy and Euler class), once taking the fiber-connected sum with Φ_m, the twistedness information disappears. We note that the existence of a Lefschetz singular fiber, which appears in Lemma 3.3 as the assumption on the monodromy around each q_i, is essential for the trick.

Remark 3.1 A T^2-bundle over S^2 is determined uniquely up to isomorphism by its Euler class, which is the obstruction class for the existence of a cross-section and lies in $H^2\left(S^2; \pi_1(T^2)\right) = \mathbb{Z}^2$. We note that the bundle with Euler class (m, n) is isomorphic to that with Euler class $(d, 0)$, where d is the greatest common divisor of m and n. Hence, the set of isomorphism classes of T^2-bundles over S^2 is not $\mathbb{Z}^2$, but $\mathbb{Z}_{\geq 0}$. The classification of T^2-bundles over T^2 was completed by Sakamoto and Fukuhara [161]. On the other hand, as far as the author knows, the classification of T^2-bundles over Σ_g with $g \geq 2$ has not yet been established. It would be an interesting problem to complete the classification.

3.4 The Matsumoto–Fukaya Fibration

As we have seen in the preceding sections, Lefschetz fibrations are useful for the classification of diffeomorphism type of elliptic surfaces. Matsumoto generalized the notion of Lefschetz fibrations as follows.

Definition 3.4 (Achiral Lefschetz Fibration) Let $f : M \to B$ be a smooth map satisfying the conditions (1), (2) of a Lefschetz fibration, and the following (3') instead of (3).

(3') For each critical point p, there exist local complex coordinates (z_1, z_2) and w around $p \in M$ and $f(p) \in B$, respectively, such that f is locally expressed as $w = f(z_1, z_2) = z_1 z_2$ or $w = f(z_1, z_2) = z_1 \bar{z}_2$.

Then f is called an *achiral Lefschetz fibration*. A critical point of f is called a *positive* (resp. *negative*) *Lefschetz singularity*, if its local description is given by $f(z_1, z_2) = z_1 z_2$ (resp. $f(z_1, z_2) = z_1 \bar{z}_2$), and the corresponding fiber is called a *positive* (resp. *negative*) *singular fiber*.

Notice that the monodromy around a negative singular fiber is a left-handed Dehn twist about the vanishing cycle, while that around a positive singular fiber is a right-handed Dehn twist. In the case where the fiber genus is 1, an achiral Lefschetz fibration is a torus fibration whose singular fiber is either of type I_1^+ or I_1^-. Matsumoto and Fukaya [118] found such a torus fibration structure on the 4-sphere S^4 in the following manner. Let $H : S^3 \to \mathbb{C}P^1$ be the Hopf fibration and $\Sigma H : S^4 \to S^3$ its suspension. Then, by taking their composition, we obtain a torus fibration $f_{MF} := H \circ \Sigma H : S^4 \to S^2$. If the two pinched points of the suspension $\Sigma S^2 = S^3$ sit on the same fiber of $H : S^3 \to \mathbb{C}P^1$, then the torus fibration has only one singular fiber with two Lefschetz singularities of opposite sign. Such a singular fiber is called a *twin*, following the terminology of Montesinos. On the other hand, if the two pinched points are on different Hopf fibers, then the fibration f_{MF} becomes an achiral Lefschetz fibration with only two singular fibers of opposite sign.

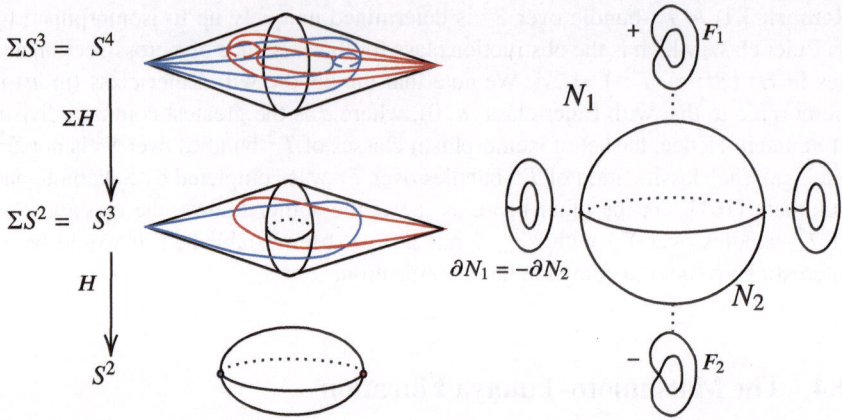

The situation is described in the above figures (the left-hand figure appears in Matsumoto's article [119]). The two pinched points of the suspension $\Sigma S^3 = S^4$ become two Lefschetz singularities of opposite sign, because the cone of the positive (resp. negative) Hopf link produces a positive (resp. negative) Lefschetz singularity in D^4, which is the cone of S^3.

Matsumoto originally found this fibration on S^4 by using Montesinos' twin, and explained his construction in a lecture at the University of Tokyo. After the lecture, Fukaya, who was a master's student at that time, suggested to him the above construction using the Hopf fibration. Based on such a background, we call $f_{MF} \colon S^4 \to S^2$ the Matsumoto–Fukaya fibration, following the proposal by Matsumoto in a private conversation.

At first glance, the torus fibration f_{MF} seems to have no relation with complex geometry, since it has a negative Lefschetz singularity and the total space S^4 has no almost complex structure. As will be revealed in Sect. 4.4, however, this example plays an central role in the construction of non-Kähler complex structures on $\mathbb{R}^4$. As a preliminary for the construction, let us analyze the structure of this fibration in detail.

Let F_1 (resp. F_2) be the only positive (resp. negative) singular fiber of f_{MF}, and N_1 (resp. N_2) its tubular neighborhood in S^4. Then, we can describe the 4-sphere S^4 as the union $N_1 \cup N_2$. Indeed, if we split the base S^2 as the union of two closed 2-disks D_1 and D_2, where $f_{MF}(F_j) \in D_j$ and $\partial D_1 = \partial D_2$, then we can take the two tubular neighborhoods by $N_1 = f_{MF}^{-1}(D_1)$ and $N_2 = f_{MF}^{-1}(D_2)$.

We denote the restriction of f_{MF} to N_j by f_j $(j = 1, 2)$. Then, $f_1 \colon N_1 \to D_1$ is a genus-one Lefschetz fibration with only one positive singularity, and $f_2 \colon N_2 \to D_2$ is a genus-one achiral Lefschetz fibration with only one negative singularity. Each monodromy is a right-handed Dehn twist and a left-handed Dehn twist about the vanishing cycle, respectively. Therefore, ∂N_1 and ∂N_2 are T^2 bundles over S^1 which are orientation-reversing diffeomorphic.

Next, we describe the gluing diffeomorphism between ∂N_1 and ∂N_2. First the vanishing cycles of two singularities coincide, so we call its representing homology class in $H_1(T^2; \mathbb{Z}) = \mathbb{Z}^2$ the meridian. Then the gluing diffeomorphism between ∂N_1 and ∂N_2 can be described as one twist along the longitude of the regular torus fiber, namely, the multiplicity-1 logarithmic transformation after the trivial fiberwise gluing. (This twist arises from the Hopf fibration. Indeed, if we collapse all the vanishing cycles of fibers of N_1 and N_2 to have two solid tori and glue them with the same twist, then the Hopf fibration $H: S^3 \to S^2$ is obtained.) By this diffeomorphism of the boundaries, we obtain $N_1 \cup N_2 = S^4$. The exact proof of this fact is given by the Kirby diagram and its calculation (see Example 8.4.7 in [61] and Figures 1, 2, 3 in [29]). Now we remove from N_2 a standard neighborhood X of the negative singularity. Since X is the total space of an annulus fibration over D_2 with one negative singularity, it is indeed the standard neighborhood model of a negative Lefschetz singularity, which is diffeomorphic to the closed 4-ball B^4. Hence, the complement $N_1 \cup (N_2 \backslash X)$ is diffeomorphic to $\mathbb{R}^4$. Since the fibration $N_2 \backslash X \to D_2$ doesn't have a singularity anymore, it is the trivial annulus bundle over D_2. Thus the total space is diffeomorphic to the product $A \times D^2$, where A denotes an annulus. We sum up these observations into the following lemma.

Lemma 3.4 *We glue the product $A \times D^2$ to N_1 as follows. For each $\theta \in \partial D^2 = -\partial D_1 \cong S^1$, the annulus $A \times \{\theta\}$ embeds as a thickened meridian of the regular fiber $f_1^{-1}(\theta) \cong T^2$ so that when $t \in S^1$ rotates once, the annulus rotates once in the longitude direction of T^2. Then the resultant manifold is diffeomorphic to $\mathbb{R}^4$.*

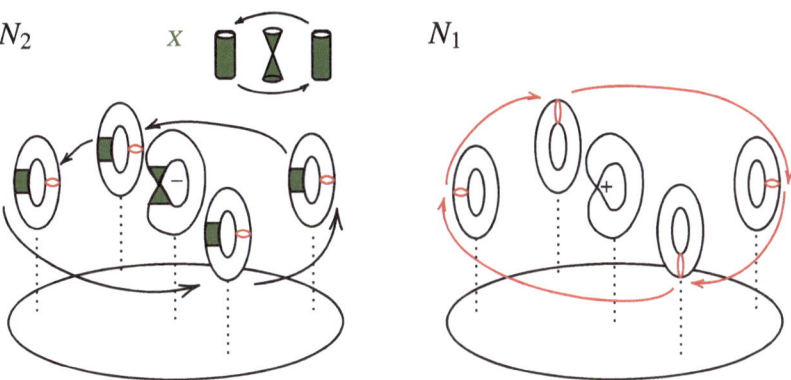

Chapter 4
Non-Kähler Complex Structures on $\mathbb{R}^{2n}$

In this chapter, we explain the construction of non-Kähler complex structures on $\mathbb{R}^{2n}$ for any $n \geq 2$. The case $n > 2$ is due to Calabi and Eckmann [17] and the case $n = 2$ is due to Di Scala, Kasuya and Zuddas [29].

4.1 The Problem and the Key Lemma

Now let us consider the following simple question.

Problem 4.1 Is there any non-Kähler complex structure on $\mathbb{R}^{2n}$?

When $n = 1$, the answer is clearly negative, since all the complex curves are Kähler. On the other hand, the answer is in fact affirmative if $n \geq 2$. In this chapter, we first introduce the construction of non-Kähler complex structures on $\mathbb{R}^{2n}$ with $n \geq 3$ as an application of the famous Calabi–Eckmann manifold, and then, explain the construction of non-Kähler complex structures on $\mathbb{R}^4$ by Di Scala, Kasuya and Zuddas.

As we have seen in Theorem 2.5, a compact complex surface is Kähler if and only if its first Betti number is even. In general, an odd-dimensional Betti number of a compact Kähler manifold is even by the Hodge theory (see Theorem 1.10). If the manifold is open, however, such a statement does not hold anymore. Indeed, it is easy to show that there exist Stein manifolds with odd first Betti number, or to prove that any open oriented 4-manifold admits a Kähler metric. Thus, for open manifolds it is impossible to distinguish Kähler and non-Kähler only by topological information. Then the following simple lemma becomes important in the non-compact case.

Lemma 4.1 *If a complex manifold contains a compact complex curve which represents the trivial second homology class, then the manifold is non-Kähler.*

N. Kasuya, *Non-Kähler Complex Surfaces and Strongly Pseudoconcave Surfaces*, SpringerBriefs in Mathematics, https://doi.org/10.1007/978-981-96-3002-8_4

Proof Let (X, J) be a Kähler manifold, and C be a compact complex curve in it. Then, for a symplectic form ω compatible with J, we have $\int_C \omega > 0$. Hence, the curve C represents a nontrivial class in $H_2(X; \mathbb{R})$. Therefore, a complex manifold containing a homologically trivial compact complex curve is not Kähler. $\qquad\square$

4.2   The Calabi–Eckmann Manifolds

In [17], Calabi and Eckmann gave complex structures on the product of two odd-dimensional spheres. Let $z = (z_0, \ldots, z_p)$ and $w = (w_0, \ldots, w_q)$ be the coordinates on $\mathbb{C}^{p+1}$ and $\mathbb{C}^{q+1}$, respectively. They constructed an explicit holomorphic elliptic bundle $S^{2p+1} \times S^{2q+1} \to \mathbb{C}P^p \times \mathbb{C}P^q$ in the following way. First, let $h_p \colon S^{2p+1} \to \mathbb{C}P^p$ and $h_q \colon S^{2q+1} \to \mathbb{C}P^q$ be the Hopf fibrations and take their direct product

$$h_{p,q} \colon S^{2p+1} \times S^{2q+1} \to \mathbb{C}P^p \times \mathbb{C}P^q.$$

Then $h_{p,q}$ is a smooth T^2-bundle. Now let $\tau \in \mathbb{C}$ be a constant with $\mathrm{Im}\,\tau > 0$, and $S(\tau)$ be the elliptic curve of modulus τ, namely,

$$S(\tau) = \mathbb{C}/\mathbb{Z} \oplus \tau\mathbb{Z}.$$

We take the standard atlas $\{U_i \times U_j\} = \{([z], [w]) \mid z_i \neq 0, w_j \neq 0\}$ ($0 \leq i \leq p$, $0 \leq j \leq q$) of the complex manifold $\mathbb{C}P^p \times \mathbb{C}P^q$. Then the map $f_{ij} \colon h_{p,q}^{-1}(U_i \times U_j) \to \mathbb{C}^p \times \mathbb{C}^q \times S(\tau)$ given by

$$f_{ij}(z, w)$$
$$= \left(\frac{1}{z_i}(z_0, \ldots, \widehat{z_i}, \ldots, z_p), \frac{1}{w_j}(w_0, \ldots, \widehat{w_j}, \ldots, w_q), \frac{1}{2\pi i}(\log z_i + \tau \log w_j) \right)$$

induces a complex structure on $h_{p,q}^{-1}(U_i \times U_j)$. If $([z], [w]) \in (U_i \times U_j) \cap (U_k \times U_l)$, the transition function $g_{ij}^{kl} \colon (U_i \times U_j) \cap (U_k \times U_l) \to S(\tau)$ of the T^2-bundle $h_{p,q}$ is given by

$$g_{ij}^{kl}([z], [w]) = \frac{1}{2\pi i}\left(\log \frac{z_k}{z_i} + \tau \log \frac{w_l}{w_j} \right),$$

which is holomorphic. Thus we obtain a complex structure on $S^{2p+1} \times S^{2q+1}$, for which $h_{p,q}$ becomes a holomorphic bundle whose fibers are all isomorphic to the elliptic curve $S(\tau)$. This is called the *Calabi–Eckmann manifold*, which we denote by $M_{p,q}(\tau)$. Notice that for different moduli τ and τ', two complex manifolds $M_{p,q}(\tau)$ and $M_{p,q}(\tau')$ are not biholomorphic to each other. Hence there exist uncountable many distinct complex structures on $S^{2p+1} \times S^{2q+1}$.

Now we can give non-Kähler complex structures on $\mathbb{R}^{2(p+q+1)}$ by taking an appropriate open subset of $M_{p,q}(\tau)$. First, we give the simplest cellular decomposition of $S^{2p+1} \times S^{2q+1}$ and then, take the top dimensional cell, which is diffeomorphic to $\mathbb{R}^{2(p+q+1)}$. We denote the open subset of $M_{p,q}(\tau)$ corresponding to that cell by $E_{p,q}(\tau)$. If p and q are positive, it contains almost all the elliptic fibers of $h_{p,q}$. Then, by Lemma 4.1, $E_{p,q}(\tau)$ gives a non-Kähler complex structure on $\mathbb{R}^{2(p+q+1)}$.

We note that when $p = 0$ and $q = 1$, the manifold $M_{0,1}(\tau)$ is just a Hopf surface, and $E_{0,1}(\tau)$ is an open subset of $\mathbb{C}^2$. Therefore, in order to settle the case $n = 2$, another strategy is needed.

4.3 Di Scala–Kasuya–Zuddas Construction

In [29], Di Scala, Zuddas and the author proved the following theorem.

Theorem 4.1 (Di Scala, Kasuya and Zuddas [29]) *For any pair of real numbers (ρ_1, ρ_2) such that $1 < \rho_2 < \rho_1^{-1}$, there exist a complex surface $E(\rho_1, \rho_2)$ and a surjective holomorphic map $f: E(\rho_1, \rho_2) \to \mathbb{C}P^1$ satisfying the following conditions:*

1. *Each $E(\rho_1, \rho_2)$ is diffeomorphic to $\mathbb{R}^4$.*
2. *The fibers of f are classified into the three types; a non-singular elliptic curve, a singular elliptic fiber of type I_1, and a holomorphic annulus.*
3. *If $(\rho_1, \rho_2) \neq (\rho_1', \rho_2')$, $E(\rho_1, \rho_2)$ and $E(\rho_1', \rho_2')$ are not biholomorphic to each other.*

In particular, there exist uncountably many non-Kähler complex structures on $\mathbb{R}^4$.

The proof is by an explicit construction. We use the topological decomposition of $\mathbb{R}^4$ into the two pieces N_1 and $A \times D^2$ given in Lemma 3.4. In other words, the Matsumoto–Fukaya fibration provides the topological design for the construction of $E(\rho_1, \rho_2)$ and f.

Suppose the real numbers ρ_0, ρ_1, ρ_2 satisfy the condition $0 < \rho_0 < \rho_1 < 1 < \rho_2 < \rho_1^{-1}$. First we induce a complex structure on each of the pieces N_1 and $A \times D^2$. The easy one is $A \times D^2$. We can take the product $\Delta(1, \rho_2) \times \Delta(\rho_0^{-1})$ of a holomorphic annulus and a holomorphic disk. On the other hand, N_1 is the total space of the genus-one Lefschetz fibration $f_1: N_1 \to D_1$. Hence, the appropriate complex manifold is the neighborhood model of a singular fiber of type I_1 in an elliptic surface (see Sect. 2.2). Let us recall the model. We first take an elliptic fibration

$$\pi : \mathbb{C}^* \times \Delta(0, \rho_1)/\mathbb{Z} \to \Delta(0, \rho_1),$$

where the action of $n \in \mathbb{Z}$ is given by

$$n \cdot (z, t) = (zt^n, t).$$

This fibration extends over $\Delta(\rho_1)$ as a singular elliptic fibration $g_1 : W \to \Delta(\rho_1)$. This is Kodaira's neighborhood model of a singular fiber of type I_1.

Now we construct the complex surface $E(\rho_1, \rho_2)$, by a complex analytic gluing of W and the product $\Delta(1, \rho_2) \times \Delta(\rho_0^{-1})$. The gluing domain in the product is given by $\Delta(1, \rho_2) \times \Delta(\rho_1^{-1}, \rho_0^{-1})$. From W, we define a gluing domain which is biholomorphic to $\Delta(1, \rho_2) \times \Delta(\rho_0, \rho_1)$ as follows. Recalling Proposition 2.3 and setting $m = 1$ and $h \equiv 0$, we obtain the multi-valued holomorphic function $\varphi :$ $\Delta(\rho_0, \rho_1) \to \mathbb{C}^*$ defined by

$$\varphi(t) = \exp\left(\frac{1}{4\pi i}(\log t)^2 - \frac{1}{2}\log t\right).$$

Then the branches are compatible with the action of $\mathbb{Z}$ to $\mathbb{C}^*$, since it satisfies

$$\varphi(re^{i(\theta+2\pi)}) = re^{i\theta}\varphi(re^{i\theta}) = t\varphi(t). \tag{4.1}$$

Hence φ defines a holomorphic section of π (as was verified in Proposition 2.3). Taking into account this property of φ, we define

$$Y := \{(z\varphi(t), t) \in \mathbb{C}^* \times \Delta(\rho_0, \rho_1) \mid z \in \Delta(1, \rho_2)\}.$$

Then Y is invariant under the action of $\mathbb{Z}$, and the quotient $V := Y/\mathbb{Z}$ is a domain in W along the holomorphic section defined by φ, which is biholomorphic to $\Delta(1, \rho_2) \times \Delta(\rho_0, \rho_1)$. This is the gluing domain in W. The biholomorphism j between the two gluing domains is given by

$$j : V \cong \Delta(1, \rho_2) \times \Delta(\rho_0, \rho_1) \to \Delta(1, \rho_2) \times \Delta(\rho_1^{-1}, \rho_0^{-1}); \ (z, t) \mapsto (z, t^{-1}).$$

Then we obtain the complex surface

$$E(\rho_1, \rho_2) := W \cup_j \left(\Delta(1, \rho_2) \times \Delta(\rho_0^{-1})\right).$$

This is diffeomorphic to $\mathbb{R}^4$ by the following reason. Since the gluing domain V is the product trivialized by φ with the fiber $\Delta(1, \rho_2)$, the condition (4.1) of φ implies that the annulus $\Delta(1, \rho_2)$ rotates once in the longitude direction of the elliptic fiber, when t rotates once around 0. For, the value of φ varies by the multiplication by t when the argument of t increases by 2π, which means that the annulus $\Delta(1, \rho_2)$ moves to the next fundamental domain of the elliptic curve $\mathbb{C}^*/\mathbb{Z}$. Notice that the argument and radial directions of $\mathbb{C}^*$ correspond to the meridian and the longitude, respectively. Therefore, the gluing of W and $\Delta(1, \rho_2) \times \Delta(\rho_0^{-1})$ is topologically the same as that of Lemma 3.4, and thus, $E(\rho_1, \rho_2)$ is diffeomorphic to $\mathbb{R}^4$.

Finally, we define f as the projection to $\Delta(\rho_1)$ and $\Delta(\rho_0^{-1})$, respectively. Namely, f is equal to g_1 on W, and to the projection to the second factor on $\Delta(1, \rho_2) \times \Delta(\rho_0^{-1})$. Since the two disks $\Delta(\rho_1)$ and $\Delta(\rho_0^{-1})$ form $\mathbb{C}P^1$ by the gluing biholomorphism

$$\Delta(\rho_0, \rho_1) \to \Delta(\rho_1^{-1}, \rho_0^{-1}); \ t \mapsto t^{-1},$$

we obtain the surjective holomorphic map $f \colon E(\rho_1, \rho_2) \to \mathbb{C}P^1$.

Remark 4.1 Notice that the parameter ρ_0 only plays a role to determine the size of the gluing domains in W and $\Delta(1, \rho_2) \times \Delta(\rho_0^{-1})$. Hence, the obtained complex structure depends only on ρ_1 and ρ_2, and not on ρ_0.

4.4 Various Properties of $E(\rho_1, \rho_2)$

The complex surface $E(\rho_1, \rho_2)$ has many interesting features. First, compact holomorphic curves in $E(\rho_1, \rho_2)$ can be easily classified.

Lemma 4.2 *Any compact holomorphic curve in $E(\rho_1, \rho_2)$ is a compact fiber of f.*

Proof Let $i \colon C \to E(\rho_1, \rho_2)$ be a compact holomorphic curve, that is, i is a holomorphic immersion of a compact Riemann surface C. What we have to prove is that the composition $f \circ i \colon C \to \mathbb{C}P^1$ is a constant map. Since $f \circ i$ is a holomorphic map between compact Riemann surfaces, it is either a branched covering map or a constant map. On the other hand, i is null-homotopic as a continuous map since $E(\rho_1, \rho_2) \cong \mathbb{R}^4$ is a contractible space. Hence, $f \circ i$ is also null-homotopic and cannot be a branched covering. Therefore, it must be a constant map. $\qquad\square$

Namely, any compact holomorphic curve in $E(\rho_1, \rho_2)$ can be expressed as $f^{-1}(t)$ $(t \in \Delta(\rho_1))$, which is the only rational curve with a node if $t = 0$, and an elliptic curve of modulus $\dfrac{1}{2\pi i} \log t$ if $t \neq 0$. Based on this classification, we can easily prove that there exist uncountably many non-Kähler complex structures on $\mathbb{R}^4$.

Proposition 4.1 (Theorem 4.1 (3)) *If $(\rho_1, \rho_2) \neq (\rho_1', \rho_2')$, $E(\rho_1, \rho_2)$ and $E(\rho_1', \rho_2')$ are not biholomorphic to each other.*

Proof We prove the contraposition. Suppose there exists a biholomorphism

$$\Phi \colon E(\rho_1, \rho_2) \to E(\rho_1', \rho_2').$$

Then it maps a compact curve in $E(\rho_1, \rho_2)$ to a compact curve in $E(\rho_1', \rho_2')$, so we have $\Phi(W) = W'$. Since an elliptic curve is mapped to the elliptic curve of the same modulus, Φ is a fiberwise biholomorphism on W and its base map $\Delta(\rho_1) \to \Delta(\rho_1')$ is the identity. Therefore, we have $\rho_1 = \rho_1'$. Moreover, by the analytic continuation, Φ is a fiberwise biholomorphism on the whole $E(\rho_1, \rho_2)$. Therefore, the two annuli $\Delta(1, \rho_2)$ and $\Delta(1, \rho_2')$ are biholomorphic, and we have $\rho_2 = \rho_2'$.   □

By a similar argument, it can be proven that any meromorphic function on $E(\rho_1, \rho_2)$ is the pullback by f of that on $\mathbb{C}P^1$ ([30], Theorem 1.3 (1)).

Next, we prove that the Picard group $\mathrm{Pic}(E(\rho_1, \rho_2))$ is uncountable. Let $O_{\mathbb{C}P^1}(k)$ be the holomorphic line bundle over $\mathbb{C}P^1$ with first Chern class k, and L_k be its pullback by f. Moreover, We denote by f^* the homomorphism between the Picard groups induced by f.

Theorem 4.2 *The homomorphism* $f^*\colon \mathrm{Pic}(\mathbb{C}P^1) \to \mathrm{Pic}(E(\rho_1, \rho_2))$ *is injective, and the Picard group* $\mathrm{Pic}(E(\rho_1, \rho_2))$ *is a non-trivial complex vector space.*

Proof We suppose that L_k is the trivial line bundle, and prove that $O_{\mathbb{C}P^1}(k)$ is the trivial line bundle. Let τ be a non-vanishing holomorphic section of L_k. On the other hand, since $O_{\mathbb{C}P^1}(k)$ is trivial over each of $\Delta(\rho_1)$ and $\Delta(\rho_0^{-1})$, there exist non-vanishing holomorphic sections σ_1 and σ_2 defined over $\Delta(\rho_1)$ and $\Delta(\rho_0^{-1})$.

By pulling back these sections by f, we obtain the non-vanishing section $f^*(\sigma_1)$ over $W_1 := W$ and the non-vanishing section $f^*(\sigma_2)$ over $W_2 := \Delta(1, \rho_2) \times \Delta(\rho_0^{-1})$. Then we obtain holomorphic functions τ_j on W_j ($j = 1, 2$) by

$$\tau|_{W_j} = \tau_j f^*(\sigma_j).$$

Since $W_1 = W$ is foliated by compact fibers, τ_1 is fiberwise constant, namely, there exists a holomorphic function u_1 on $\Delta(\rho_1)$ such that $\tau_1 = f^*(u_1)$. On the intersection $V = W_1 \cap W_2$, we have

$$f^*(u_1\sigma_1) = \tau_2 f^*(\sigma_2),$$

and hence, τ_2 is fiberwise constant on V. By the analytic continuation, τ_2 is fiberwise constant on the whole W_2. Namely, there exists a holomorphic function u_2 on $\Delta(\rho_0^{-1})$ such that $\tau_2 = f^*(u_2)$. Then, $u_1\sigma_1$ and $u_2\sigma_2$ defines a non-vanishing holomorphic section on $O_{\mathbb{C}P^1}(k)$, and hence, $O_{\mathbb{C}P^1}(k)$ is trivial. This proves the injectivity of f^*.

Since f^* is injective, $\mathrm{Pic}(E(\rho_1, \rho_2))$ contains $\mathrm{Pic}(\mathbb{C}P^1) = \mathbb{Z}$ as a subgroup. Moreover, by the long exact sequence of sheaf cohomology, we have

$$\mathrm{Pic}(E(\rho_1, \rho_2)) = H^1(E(\rho_1, \rho_2), O^*) \cong H^1(E(\rho_1, \rho_2), O).$$

Therefore, $\mathrm{Pic}(E(\rho_1, \rho_2))$ is a non-trivial complex vector space.   □

A similar argument proves that holomorphic vector bundles $L_{k_1} \oplus L_{k_2} \oplus \cdots \oplus L_{k_n}$ ($k_1 \leq k_2 \leq \cdots \leq k_n$) on $E(\rho_1, \rho_2)$ are pairwise non-isomorphic. By taking the total space, we obtain non-Kähler complex structures on $\mathbb{R}^{2n+4}$ that are pairwise non-biholomorphic. They are obviously different from the complex structures constructed by Calabi and Eckmann by the classification of compact complex curves ([30], Theorem 4).

Finally, we show the following remarkable fact.

Theorem 4.3 *Any connected open orientable* 4-*manifold* M^4 *admits uncountably many non-Kähler complex structures.*

For the proof, Phillips' theorem [154] is essential.

Theorem 4.4 (Phillips [154]) *Let* M *be an open manifold. Then the map*

$$d: \mathrm{Sub}(M, V) \to \mathrm{Epi}(TM, TV); \ f \mapsto df$$

is a weak homotopy equivalence, where $\mathrm{Sub}(M, V)$ *is the space of submersions from* M *to* V, *and* $\mathrm{Epi}(TM, TV)$ *is the space of surjective homomorphisms from* TM *to* TV.

Corollary 4.1 *Let* M *be an open parallelizable manifold. Then there exists a submersion from* M *to* $\mathbb{R}^n$ *if* $n \leq \dim M$.

A smooth orientable manifold M is said to be *spin* if $w_2(M) = 0$, where $w_2(M) \in H^2(M; \mathbb{Z}_2)$ denotes the 2nd Stiefel–Whitney class. Recalling that $w_2(M)$ is the obstruction to the existence of a tangential trivialization over the 2-skeleton of a cell decomposition of M, we can easily prove that an open, spin 4-manifold M^4 is parallelizable. Hence, if M^4 is open and spin, then there exists an immersion $g: M^4 \to \mathbb{C}^2$ by Corollary 4.1. Then M^4 admits a Kähler complex structure, since we can pull back the complex structure on $\mathbb{C}^2$ by g. In the case where M^4 is a general connected open orientable 4-manifold, the same strategy works if we change the target space to $\mathbb{C}P^2$ instead of $\mathbb{C}^2$. Now we are ready to prove Theorem 4.3.

Proof of Theorem 4.3 First we give the proof for the case where M^4 is spin (so, it is parallelizable). By Corollary 4.1, there exists an immersion $h: M^4 \to E(\rho_1, \rho_2)$. Since h is an immersion, there exists a small 4-ball $B \subset M^4$ such that the restriction of h to B is an embedding. Now we take positive numbers ρ_1' and ρ_2' such that $0 < \rho_1' < \rho_1$ and $1 < \rho_2' < \rho_2$. Then we have $E(\rho_1', \rho_2') \subset E(\rho_1, \rho_2)$, and it is topologically an open 4-ball embedded in $\mathbb{R}^4$. Since $h(B)$ is also an open 4-ball in $E(\rho_1, \rho_2) \cong \mathbb{R}^4$, we can take a diffeomorphism of $\mathbb{R}^4$ that maps $h(B)$ to $E(\rho_1', \rho_2')$. Rescaling h by composing with this diffeomorphism, we may assume that $h(B) = E(\rho_1', \rho_2')$. Pulling back the complex structure on $E(\rho_1, \rho_2)$ by h, M^4 becomes a complex surface in which $E(\rho_1', \rho_2')$ is holomorphically embedded. Therefore, M^4 admits a non-Kähler complex structure. Moreover, it is easily proven that there are uncountably many non-Kähler complex structures by changing (ρ_1, ρ_2) and (ρ_1', ρ_2'). Even in the case where M^4 is non-spin, we can do a similar argument by changing $E(\rho_1, \rho_2)$ to its one point blow-up. The point is that $E(\rho_1, \rho_2)\#\overline{\mathbb{C}P^2}$ is diffeomorphic to $\overline{\mathbb{C}P^2} - \{\mathrm{pt}\}$ and the restriction of its tangent bundle to the (-1)-curve is isomorphic to the unique non-trivial rank 4 real vector bundle ξ over S^2. On the other hand, M^4 admits a non-vanishing vector field and an almost complex structure. Hence, the tangent bundle TM^4, regarded as rank 2 complex vector bundle, splits into the Whitney sum of a complex line bundle η and the trivial complex line bundle ε^1. Now it is an easy exercise to show that there exists a bundle epimorphism from $TM^4 \cong \eta \oplus \varepsilon^1$ to ξ. □

Part II
Strong Pseudoconvexity, Pseudoconcavity and Contact Structures

Chapter 5
Strongly Pseudoconvex Manifolds

In this chapter, we review strongly pseudoconvex manifolds, the central object in the theory of functions of several complex variables. First we recall the definition of pseudoconvexity in Sect. 5.1, give a brief review of the facts about Stein manifolds and Stein spaces in Sect. 5.2, and then, in Sect. 5.3, introduce Eliashberg's handlebody construction of Stein manifolds. In Sects. 5.4 and 5.5, we summarize the known results about strongly pseudoconvex CR manifolds from the viewpoint of fillability (=embeddability) and its variations. For the topics of this chapter, [67] and [49] are good references, and a good part of Sects. 5.4 and 5.5 is based on Ohsawa's survey [144]. For the L^2-theory, another important approach in analysis of several complex variables which we couldn't explain in this book, the reader is referred to [143].

5.1 Pseudoconvexity

It is well-known that Oka's affirmative solution to the following problem has contributed much to the development of the theory of functions of several complex variables.

Problem 5.1 (Levi Problem) Is a pseudoconvex domain a domain of holomorphy?

A domain $D \subset \mathbb{C}^n$ is called a *domain of holomorphy* if there exists a holomorphic function on D which does not admit the analytic continuation to any domain across the boundary ∂D. A *defining function* for a domain $D \subset \mathbb{C}^n$ with C^2 smooth boundary is a C^2 smooth function $\varphi \colon \mathbb{C}^n \to \mathbb{R}$ such that $D = \{z \in \mathbb{C}^n \mid \varphi(z) < 0\}$ and $d\varphi_z \neq 0$ for every $z \in \partial D$. A *pseudoconvex domain* is a domain in $\mathbb{C}^n$ admitting a defining function that is a plurisubharmonic function, which is defined as follows.

N. Kasuya, *Non-Kähler Complex Surfaces and Strongly Pseudoconcave Surfaces*, SpringerBriefs in Mathematics, https://doi.org/10.1007/978-981-96-3002-8_5

Definition 5.1 (Plurisubharmonic Function)   An upper semi-continuous function φ on a complex n-manifold X is said to be (*strictly*) *plurisubharmonic* if it is (strictly) subharmonic on any holomorphic curve in X. When φ is a C^2 smooth function, this condition is equivalent to that all the eigenvalues of its complex Hessian

$$\left(\frac{\partial^2 \varphi}{\partial z_j \partial \bar{z}_k} \right)_{1 \le j, k \le n}$$

are non-negative (resp. positive) at each point $p \in X$, where $z = (z_1, \ldots, z_n)$ are local coordinates around p. It is also equivalent to the condition that the Levi form $i \partial \bar{\partial} \varphi$ is non-negative (resp. positive) everywhere.

In general, a complex manifold admitting a defining function that is plurisubharmonic is called a pseudoconvex manifold. To be more exact, we define it as follows. A compact complex manifold X with smooth boundary is called a (*strongly*) *pseudoconvex manifold* if there exists a collar neighborhood U of the boundary ∂X and a (strictly) plurisubharmonic function $\varphi \colon U \to \mathbb{R}$ without critical points such that $\partial X = \varphi^{-1}(0)$ and $U = \{\varphi \le 0\}$. In this case, the boundary ∂X is called the (strongly) pseudoconvex boundary, and the function φ is called a *defining function* of X.

The most important example of a strongly pseudoconvex manifold is a Stein manifold, which we introduce below. The complex manifold obtained from the resolution of normal singularities of a Stein space is also strongly pseudoconvex.

5.2   Stein Spaces and Stein Manifolds

First we recall the concept of holomorphical convexity.

Definition 5.2 (Holomorphical Convexity)   Let X be a complex manifold and K its compact subset. The subset $\widehat{K} \subset X$ defined by

$$\widehat{K} = \left\{ p \in X \mid |f(p)| \le \max_{x \in K} |f(x)| \text{ for all } f \in O(X) \right\}$$

is called the *holomorphically convex hull* of K. If $\widehat{K}$ is compact for any compact subset $K \subset X$, the complex manifold X is said to be *holomorphically convex*.

Cartan and Thullen [18] have proven that a domain $D \subset \mathbb{C}^n$ is a domain of holomorphy if and only if it is holomorphically convex. Based on this concept, a Stein manifold was originally defined by Karl Stein. The motivation was to generalize Oka's topological criterion for the solvability of an analytic problem (Cousin's multiplicative problem) on the domains of holomorphy.

Definition 5.3 (Stein Manifold)   A holomorphically convex complex n-manifold X is called a *Stein manifold* if the following conditions hold:

(1) For any two points x, $y \in X$ with $x \neq y$, there exists a holomorphic function f on X such that $f(x) \neq f(y)$.
(2) For any point $p \in X$, there exist holomorphic functions $f_1, \ldots, f_n$ on X which give a local coordinate system around p.

There are two important characterizations of Stein manifolds.

Theorem 5.1 (Remmert [156], Bishop [10], Narasimhan [137])   *A Stein manifold of complex dimension n can be embedded in $\mathbb{C}^{2n+1}$ as a closed complex submanifold.*

Theorem 5.2 (Grauert [62])   *A complex manifold which admits an exhausting strictly plurisubharmonic function is Stein.*

The above theorems indeed give characterizations of Stein manifolds, because a closed complex submanifold embedded in $\mathbb{C}^N$ is clearly Stein, and it also admits an exhausting strictly plurisubharmonic function. These properties are sometimes referred as the definition of Stein manifolds.

Remark 5.1   Theorem 5.1 was first proven by Remmert under the assumption that the target dimension N is large enough, and later, Bishop and Narasimhan independently showed that it is enough to assume $N = 2n+1$. After that, Eliashberg and Gromov [39] proved that the theorem also holds with $N = [\frac{3n}{2}] + 1$ if $n \geq 2$.

Next, we see the definition of a Stein space, which is also an important subject in the theory of functions of several complex variables. In order for that, we need the notion of complex spaces instead of complex manifolds. A *complex subvariety of $\mathbb{C}^n$* is the common vanishing locus of a finite set of holomorphic functions. A *complex space* is the space obtained by gluing pieces of complex subvarieties of Euclidean spaces using biholomorphic transition maps.

Definition 5.4 (Stein Space)   A holomorphically convex complex space X is called a *Stein space* if it satisfies the condition (1) in Definition 5.3.

The next theorem gives a characterization of strongly pseudoconvex manifolds by using Stein spaces.

Theorem 5.3 (Remmert Reduction)   *For any strongly pseudoconvex manifold V, there exist a Stein space V' with finite number of normal singularities $p_1, \ldots, p_m$ and a holomorphic map $f : V \to V'$ such that the restricted map*

$$f|_{V - f^{-1}(\{p_1, \ldots, p_m\})} : V - f^{-1}(\{p_1, \ldots, p_m\}) \to V' - \{p_1, \ldots, p_m\}$$

is a biholomorphism.

We close this section by giving a typical example of a Stein manifold, which is called the Milnor fiber. Let $f(z_1, \ldots, z_n)$ be a polynomial of n-complex variables

$z_1, \ldots, z_n$ with only one critical point at the origin $\mathbf{0} \in \mathbb{C}^n$. Then $f^{-1}(0)$ is an affine variety with a singularity at $\mathbf{0}$, and for a sufficiently small positive number ε, it intersects with S_ε^{2n-1} transversely. The intersection $L = f^{-1}(0) \cap S_\varepsilon^{2n-1}$ is a closed $(2n-3)$-manifold, which is called the *link* of the singularity. In [125], Milnor proved the following.

Theorem 5.4 (Milnor) *For a sufficiently small positive number ε, the map*

$$\frac{f}{|f|} : S_\varepsilon^{2n-1} \setminus L \to S^1$$

is a fiber bundle over the circle. Moreover, there is a positive number δ such that if $0 \le |t| \le \delta$, then $V(t) := f^{-1}(t)$ transversely intersects with S_ε^{2n-1}. For such ε and δ, the map

$$f|_{f^{-1}(S_\delta^1) \cap D_\varepsilon^{2n}} : f^{-1}(S_\delta^1) \cap D_\varepsilon^{2n} \to S_\delta^1$$

is also a fiber bundle over the circle, which is isomorphic to $\frac{f}{|f|}$ as a fiber bundle.

Both fiber bundles are called the *Milnor fibration* of f. In this book, however, we only consider the latter one as the Milnor fibration. Moreover, we call each fiber

$$F_\theta = f^{-1}(\delta e^{i\theta}) \cap D_\varepsilon^{2n}$$

the *Milnor fiber* of f.

Example 5.1 The singular algebraic variety $V_0 = f^{-1}(0)$ is a Stein space, and its resolution is a strongly pseudoconvex manifold. Moreover, the Milnor fiber of f is a Stein manifold, since it is a non-singular level set of the polynomial f.

5.3 Eliashberg's Construction of Stein Manifolds

In this section, we review Eliashberg's handlebody construction of Stein manifolds (see [24, 35] for the details). Before that, let us recall a basic fact on handle decomposition of a Stein manifold. For the Morse theory and handle decompositions of smooth manifolds, see [124]. Let (X, J) be a Stein manifold, where J is the complex structure on X. (We use this notation, because here we want to regard a complex manifold as the pair of a smooth manifold and a complex structure on it.) Then, by Theorem 5.2, there exists a strictly plurisubharmonic function φ on X. By a C^2-small perturbation, if necessary, we may assume that φ is a Morse function. This yields a handle decomposition of X, the indices of whose handles satisfy the following property.

Proposition 5.1 *A Stein manifold of complex dimension n admits a handle decomposition such that the index of each handle is not greater than n.*

Proof Let (X, J) be a Stein manifold of complex dimension n and φ a strictly plurisubharmonic Morse function on it. Suppose that φ has a critical point p of index m with $m \geq n + 1$. Then the tangent space $T_p X = \mathbb{R}^{2n}$ at p contains an m-dimensional real subspace W on which the Hessian matrix of φ is negative definite. The space W contains a complex line L, since $J(W) \cap W$ is a subspace of positive even dimension. Take a holomorphic curve $c(t)$ in X with $c(0) = p$ and $c'(0) \in L$, where t varies on a small complex disk $\Delta(\rho)$. Then the real function $\varphi \circ c$ has the maximum at the origin of $\Delta(\rho)$, since it is a critical point of the function whose Hessian is negative definite. This is a contradiction, since φ is strictly plurisubharmonic, and hence, it satisfies the maximum principle. Therefore, φ has no critical point of index greater than n. $\square$

On the other hand, Eliashberg proved the converse in the case where $n > 2$. That is, he showed by a handlebody construction that if a $2n$-dimensional manifold admits an almost complex structure and has a handle decomposition such that the index of each handle is not greater than n, then it admits a Stein structure. In the case where $n = 2$, the same construction is valid, though the attaching framing of each handle is strictly restricted. Namely, the topological characterization of Stein manifolds has already been completed by the following result.

Theorem 5.5 (Eliashberg [35], Gompf [60]) *For $n > 2$, a smooth open oriented $2n$-manifold admits a Stein structure if and only if it admits an almost complex structure and has a handle decomposition without handles of index greater than n. Moreover, a smooth open oriented 4-manifold admits a Stein structure if and only if it has a handle decomposition which satisfies the following conditions:*

1. *The index of each handle is equal or smaller than 2.*
2. *Each 2-handle is attached along a Legendrian knot in the contact structure (see Sect. 6.2 for the definition) on the boundary of the handlebody consisting of all the 0-handles and 1-handles.*
3. *The attaching framing of each 2-handle is different from the contact framing of the Legendrian knot by one negative twist.*

Remark 5.2 This handle decomposition of a Stein surface is translated by Akbulut and Ozbagci [1] into the so-called positive allowable Lefschetz fibration (PALF). See also Loi and Piergallini [115].

Now we overview the handlebody construction of Eliashberg and Gompf focusing only the case $n = 2$. First we need to complex analytically attach a holomorphic handle to the strongly pseudoconvex boundary of a given Stein surface. Then we extend the strictly plurisubharmonic function on the Stein surface over the handle. In order to do so, we need to prepare the model of a holomorphic handle in $\mathbb{C}^2$, and foliate it by strongly pseudoconvex hypersurfaces of a special type.

Let $\mathbb{C}^2 = \mathbb{R}^2 \oplus i\mathbb{R}^2$ be the decomposition of $\mathbb{C}^2$ into the real and the imaginary parts, and $x = (x_1, x_2)$, $y = (y_1, y_2)$ be the coordinates on each factor, respectively.

For a positive number ϵ, we define the standard holomorphic handle of index 2 by

$$H_\epsilon^{(2)} = \left\{ x + iy \in \mathbb{C}^2 \mid \sqrt{x_1^2 + x_2^2} \le \epsilon, \sqrt{y_1^2 + y_2^2} \le 1 + \epsilon \right\}.$$

We set $D_r = \left\{ x_1 = x_2 = 0, \sqrt{y_1^2 + y_2^2} \le r^2 \right\}$ and $S = \partial D_1$. Notice that $D_{1+\epsilon}$ is the core of H_ϵ, which is a totally real disk, and S is the attaching circle. The standard handle of index 1 is also given by $H_\epsilon^{(1)} = \{|x_1| \le \epsilon, |x_2| \le 1 + \epsilon, |y_1| < \epsilon, |y_2| < \epsilon\}$. However, we omit the explanation of 1-handle attaching, since the process of 2-handle attaching is more important. Accordingly, we denote the standard 2-handle simply by H_ϵ in the following.

Now we make this handle foliated by strongly pseudoconvex hypersurfaces that can be described in the form $|y|^2 = \varphi(|x|^2)$ as drawn in the figure below. Notice that the foliation on the handle is extended over the region $\left\{ x + iy \in \mathbb{C}^2 \mid |y|^2 - a|x|^2 > 1 \right\}$ for some a with $a > 1$, as the foliation given by the level sets of the function $|y|^2 - a|x|^2$.

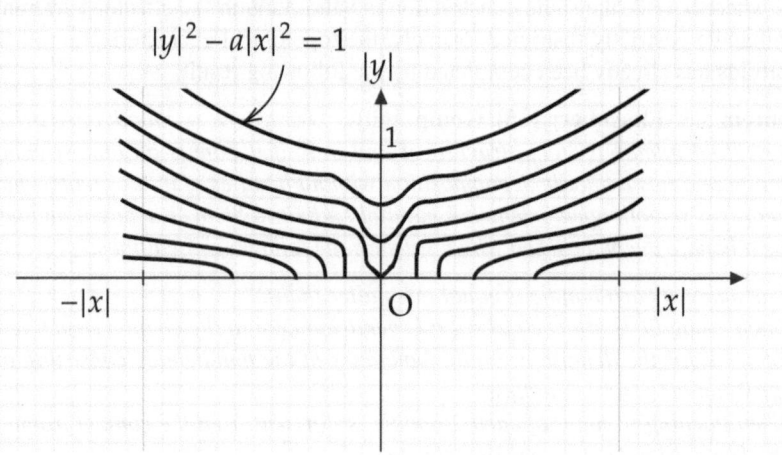

The following lemma is essential for this construction.

Lemma 5.1 Let $\Omega = \left\{ x + iy \in \mathbb{C}^2 \mid |y|^2 < \varphi(|x|^2) \right\}$. Then the hypersurface $\partial \Omega$ is strongly pseudoconvex if and only if the following two conditions hold:

$$\varphi'(t) < 1 \quad \text{and} \quad 2t\varphi(t)\varphi''(t) < (1 - \varphi'(t))(t\varphi'(t)^2 + \varphi(t)).$$

This lemma is equivalent to Proposition 4.10 in [24], but we follow Forstnerič–Kozak's expression of the conditions in Proposition 2.1 in [50]. In the original paper [35], Eliashberg gave in Lemma 3.3.1 a sufficient condition for a hypersurface of this type to be strongly pseudoconvex, and constructed the foliation of strongly pseudoconvex hypersurfaces by using the condition. Then, by the following lemma, we obtain a strictly plurisubharmonic function on the handle whose level sets coincide with the leaves of the above foliation.

Lemma 5.2 *Let $\varphi\colon V \to \mathbb{R}$ be a proper smooth regular function on a complex manifold V such that all the level sets of φ are strongly pseudoconvex. Then there exists a smooth convex increasing function $g\colon \varphi(V) \to \mathbb{R}$ such that $g \circ \varphi$ is strictly plurisubharmonic on V.*

Now we are ready to start the handle attaching. An n-dimensional submanifold M of a complex n-manifold (V, J) is said to be *totally real* if $TM \cap J(TM) = \{0\}$. A totally real submanifold M in (V, J) is said to be *J-orthogonal* to a real hypersurface $\Sigma \subset V$ if M transversely intersects with Σ and $J(T_pM) \subset T_p\Sigma$ for each intersection point p. Let (X, J) be a Stein surface with real analytic boundary ∂X and L a Legendrian knot in the contact structure $\xi = T\partial X \cap JT\partial X$. First L can be C^∞ smoothly approximated by a real analytic Legendrian knot $\widetilde{L}$ (Lemma 2.5.1 in [35], Corollary 6.25 in [24]). Then the standard handle H_ϵ can be holomorphically attached to X along $\widetilde{L}$ so that $D_{1+\epsilon}$ is J-orthogonal to ∂X. This attaching is possible, since there exist a neighborhood A of S in $D_{1+\epsilon}$ and a real analytic embedding $\psi\colon A \to X$ such that $\psi(S) = \widetilde{L}$ and $\psi(A)$ is totally real and J-orthogonal to ∂X. By taking the complexification of ψ, it extends as a biholomorphism on a small neighborhood of $A \subset H_\epsilon$. Using this biholomorphism, W and H_ϵ are holomorphically attached.

Notice that the attaching framing of the handle must be different from the contact framing by one negative twist, since the contact framing of S in the standard contact 3-sphere differs from the surface framing of D_1 by one twist.

The final step is to implant the family of strongly pseudoconvex hypersurfaces prepared in the model handle. The problem here is that the original strictly plurisubharmonic function equipped on X cannot be necessarily described as a function of $|x|$ and $|y|$ on the attaching region $X \cap H_\epsilon$. In other words, ∂X and the foliation prepared on H_ϵ don't necessarily match up. Therefore, we need to deform the original function to match up with the model (Theorem 8.5 in [24]). At this stage, the key point is that two strictly plurisubharmonic functions which coincide together with differentials along a totally real submanifold can be smoothly connected to become a new strictly plurisubharmonic function. This is stated as Lemma 3.2.2 in the original paper [35] (Proposition 3.26 in [24]), whose proof needs the following result by Richberg.

Theorem 5.6 (Richberg [157]) *Let f and g be two strictly plurisubharmonic functions defined on an open subset of $\mathbb{C}^n$. Then there exists a smooth strictly plurisubharmonic function h approximating $\max(f, g)$ from above. Moreover, h can be taken so that $h = \max(f, g)$ outside a small neighborhood of the zero set of $f - g$.*

Thus, the complex analytic attaching of the handle H_ϵ and the extension of a strictly plurisubharmonic function over it are done, which completes the proof of Theorem 5.5.

5.4 Strongly Pseudoconvex CR Manifolds (Fillability and Embeddability)

Definition 5.5 (Strongly Pseudoconvex CR Manifolds) Let M be a smooth $(2n - 1)$-manifold. A subbundle D of the complexified tangent bundle $\mathbb{C}TM = TM \otimes \mathbb{C}$ is said to be a *CR structure* on M if it satisfies the following conditions:

1. $D \cap \overline{D} = \{0\}$,
2. $[D, D] \subset D$ (this condition is called the *integrability*).

In this case, we call (M, D) a *CR $(2n - 1)$-manifold* and $\text{rank}_\mathbb{C} D$ its *CR dimension*. Moreover, a CR $(2n - 1)$-manifold (M, D) of CR dimension $n - 1$ is said to be *strongly pseudoconvex* if there exists a 1-form θ on M such that $\theta(Z) = 0$ and $id\theta(Z, \bar{Z}) > 0$ for any non-zero $Z \in D$. For such a 1-form θ, the real 1-form $i(\theta - \bar{\theta})$ is called a *contact form* of (M, D). The contact structure defined by $i(\theta - \bar{\theta})$ is called the *underlying contact structure* of (M, D). (See Chap. 6 for contact structures.)

Example 5.2 A real hypersurface Σ in a complex n-manifold (X, J) is a typical example of a CR $(2n - 1)$-manifold, where D is given by

$$D = \{X - iJ(X) \mid X \in T\Sigma \cap J(T\Sigma)\}.$$

Moreover, Σ is strongly pseudoconvex CR manifold if it is a non-singular level set of a strictly plurisubharmonic function φ, since we can take $\theta = \bar{\partial}\varphi$. Indeed, for every non-zero $Z \in D$, we have $id\theta(Z, \bar{Z}) = i\partial\bar{\partial}\varphi(Z, \bar{Z}) + i\bar{\partial}^2\varphi(Z, \bar{Z}) = i\partial\bar{\partial}\varphi(Z, \bar{Z}) > 0$.

A compact strongly pseudoconvex CR manifold is called *fillable* if it can be realized as the boundary of a Stein space. It is known that this condition is equivalent to that the CR manifold can be embedded in some $\mathbb{C}^N$ by a CR mapping (Grauert [62], Kohn [103], Rossi [158]). For higher-dimensional CR manifolds, the following striking result is known.

Theorem 5.7 (Boutet de Monvel [15]) *Any compact strongly pseudoconvex CR $(2n - 1)$-manifold is embeddable in some $\mathbb{C}^N$ if $n > 2$.*

Then, the next problem is whether any compact strongly pseudoconvex CR 3-manifold is fillable. Rossi [158] answered to this question negatively by giving the following counterexample.

Example 5.3 (Rossi Sphere) Let $f(z_1, z_2, z_3) = z_1^2 + z_2^2 + z_3^2$, where (z_1, z_2, z_3) are the coordinates on $\mathbb{C}^3$. Then, for a sufficiently small number $t \geq 0$, the intersection $L_t = f^{-1}(t) \cap S^5$ is a strongly pseudoconvex CR 3-manifold which is diffeomorphic to $\mathbb{R}P^3$. Hence we can denote it by $(\mathbb{R}P^3, J_t)$. Pulling back the structure J_t by the natural quotient map $\pi: S^3 \to \mathbb{R}P^3$, it induces a CR structure $\tilde{J}_t$ on S^3. Then, the CR manifold $(S^3, \tilde{J}_0)$ is fillable since it is the standard CR 3-sphere, while its deformation $(S^3, \widetilde{J_t})$ is not fillable if $t \neq 0$ (see [63] and Example 5.63 in [24]). This non-fillable example is called the *Rossi sphere*.

Nirenberg showed that such a phenomenon occurs quite often in dimension 3.

Theorem 5.8 (Nirenberg [140, 141]) *A generic compact strongly pseudoconvex CR 3-manifold is non-fillable.*

However, it is still an important issue to comprehend when a compact CR 3-manifold becomes fillable (embeddable), and there have been a lot of important results from the viewpoints of the closedness of the range of the tangential Cauchy-Riemann operator, the CR-Yamabe invariant, and the CR-Paneitz operator (Kohn and Rossi [105], Kohn [104], Lempert [109], Burns and Epstein [16], Epstein and Henkin [41], Chanillo, Chiu and Yang [20, 21], Takeuchi [169]).

5.5 Holomorphic Filling, Stein Filling, Concave Holomorphic Filling

For strongly pseudoconvex CR manifolds, there are several variants of fillability.

Definition 5.6 A CR manifold is said to be *holomorphically fillable* if it is realized as the boundary of a strongly pseudoconvex manifold, and *Stein fillable* if it is realized as that of a Stein domain.

Here again, the singularity link and the Milnor fibration give typical examples of a filling, a holomorphic filling and a Stein filling (recall Example 5.1).

Example 5.4 Let $f(z_1, \ldots, z_n)$ be a complex polynomial of n-variables with only one critical point at the origin $\mathbf{0}$ of $\mathbb{C}^n$. Then the singularity link $L = f^{-1}(0) \cap S_\varepsilon^{2n-1}$ $(0 < \varepsilon \ll 1)$ is naturally a strongly pseudoconvex CR $(2n - 3)$-manifold, since $f^{-1}(0)$ is a complex variety and $|z_1|^2 + \cdots + |z_n|^2$ is the standard strictly plurisubharmonic function. In this case, $f^{-1}(0)$ is a filling of L, and its resolution is a holomorphic filling of L. Moreover, the Milnor fiber $f^{-1}(t) \cap D_\varepsilon^{2n}$ $(0 < |t| \ll \varepsilon)$ is a Stein filling of the deformed CR manifold $L_t = f^{-1}(t) \cap S_\varepsilon^{2n-1}$.

Now we will organize the relationships between the three types of fillabilities. First, the holomorphical fillability and the fillability are equivalent. Indeed, by Theorem 5.3, the holomorphical fillability implies the fillability, and the converse is proved by Narasimhan [138], Hironaka [69] and the following result of Lempert.

Theorem 5.9 (Lempert [110]) *For any Stein space with only isolated singularities, its relatively compact open subsets can be biholomorphically embedded into a domain of an affine algebraic variety of the same dimension.*

On the other hand, the fillability does not imply the Stein fillability, while the converse is true by definition. However, the following result is known for the case where the manifold is 3-dimensional.

Theorem 5.10 (Bogomolov and de Oliveira [13])   *Let (V, J) be a minimal strongly pseudoconvex surface. Then there exists a deformation $\{J_t\}$ of the complex structure $J_0 = J$ such that (V, J_t) is a Stein surface for any sufficiently small positive number t.*

That is, a holomorphically fillable CR 3-manifold can be deformed into a Stein fillable CR 3-manifold. This implies that for a closed contact 3-manifold, the holomorphical fillability and the Stein fillability are equivalent. Next, we define a concave holomorphic filling of a CR manifold.

Definition 5.7   A CR manifold is said to be *concavely holomorphically fillable* if it can be realized as the boundary of a strongly pseudoconcave manifold. In this case, the strongly pseudoconcave manifold is called a *concave holomorphic filling* of the CR manifold.

Here we say that a compact complex manifold X with boundary is *strongly pseudoconcave* if there exists a defining function φ near the boundary ∂X such that $-\varphi$ is strictly plurisubharmonic. In this book, we use the terminology "concave holomorphic filling" to avoid the confusion with concave symplectic filling defined later in Chap. 6, but when there is no concern about confusion, we simply call it a concave filling. Kirimidjian proved the following interesting result on concave fillings.

Theorem 5.11 (Kiremidjian [92, 93]) *A small deformation of a concavely holomorphically fillable CR structure is concavely holomorphically fillable.*

Moreover, it follows from Theorems 5.3 and 5.9 that if a strongly pseudoconvex CR manifold is the boundary of a strongly pseudoconvex manifold, then it can be also realized as the boundary of a strongly pseudoconcave manifold. In other words, the fillability implies the concave fillability. Since the assumption of fillability is a strong condition in the 3-dimensional case, while it is automatically fulfilled when the manifold dimension is equal or greater than 5, whether the assumption can be weakened is of interest.

Problem 5.2 Which strongly pseudoconvex CR 3-manifold admits a concave holomorphic filling?

Later in Chap. 7, we will give a partial answer to Problem 5.2 (Theorem 7.1) from the viewpoint of contact structures.

Chapter 6
Contact Structures

In this chapter, as a preliminary for Chap. 7, we review the basic facts and known results about contact structures, following [53] and [151]. If the reader reads Japanese, [129] is also a good reference.

6.1 Definitions and Basic Facts

Definition 6.1 (Contact Manifold) Let N be a $(2n + 1)$-dimensional manifold. A hyperplane field η on N is a *contact structure* if it is locally described as the kernel of non-singular 1-form α satisfying $\alpha \wedge (d\alpha)^n \neq 0$. In this case, (N, η) is called a *contact manifold*, and α a *contact form*. If there exists a contact form α such that $\eta = \ker \alpha$ globally defined on N, then (N, η) is said to be *coorientable*. Moreover, suppose that n is odd and N is oriented. If the orientation coincides with that defined by the volume form $\alpha \wedge (d\alpha)^n$, we say that η is a *positive contact structure*, and otherwise, we say that η is a *negative contact structure*. For a contact form α, there exists a unique vector field R_α such that $\alpha(R_\alpha) = 1$ and $i_{R_\alpha} d\alpha = 0$. This is called the *Reeb vector field* of α.

Definition 6.2 (Contactomorphism) Let (N_1, η_1) and (N_2, η_2) be contact $(2n + 1)$-manifolds. We say that (N_1, η_1) and (N_2, η_2) are *contactomorphic* if there exists a diffeomorphism $f : N_1 \to N_2$ such that $f_*(\eta_1) = \eta_2$. In this case, f is called a *contactomorphism*.

We usually deal with contact structures up to contactomorphisms. Namely, we focus on the equivalence classes of contact structures by contactomorphisms. Here are two typical contact manifolds, both of which are called the standard contact structures.

© The Editor(s) (if applicable) and The Author(s), under exclusive license to Springer Nature Singapore Pte Ltd. 2025
N. Kasuya, *Non-Kähler Complex Surfaces and Strongly Pseudoconcave Surfaces*,
SpringerBriefs in Mathematics, https://doi.org/10.1007/978-981-96-3002-8_6

Example 6.1 (Standard Contact Structure on $\mathbb{R}^{2n+1}$) Let $(x_1, y_1, \ldots, x_n, y_n, z)$ be the Cartesian coordinates on $\mathbb{R}^{2n+1}$. Set

$$\alpha_1 = dz + \sum_{j=1}^{n}(x_j dy_j - y_j dx_j), \quad \alpha_2 = dz + \sum_{j=1}^{n} x_j dy_j.$$

Notice that if we use the polar coordinates (r_j, θ_j), α_1 can be described as

$$\alpha_1 = dz + \sum_{j=1}^{n} r_j^2 d\theta_j.$$

Then, α_1, α_2 are both contact forms. Their Reeb vector fields coincide, which is given by $R_{\alpha_1} = R_{\alpha_2} = \frac{\partial}{\partial z}$. The two contact manifolds $(\mathbb{R}^{2n+1}, \xi_1 = \ker \alpha_1)$ and $(\mathbb{R}^{2n+1}, \xi_2 = \ker \alpha_2)$ are contactomorphic via the map $f : (\mathbb{R}^{2n+1}, \xi_1) \to (\mathbb{R}^{2n+1}, \xi_2)$ defined by

$$f(x_1, y_1, \ldots, x_n, y_n, z) = \left(\sqrt{2}x_1, \sqrt{2}y_1, \ldots, \sqrt{2}x_n, \sqrt{2}y_n, z - \sum_{j=1}^{n} x_j y_j \right).$$

This contact structure is called the *standard contact structure on* $\mathbb{R}^{2n+1}$.

Example 6.2 (Standard Contact Structure on S^{2n+1})   Let $(x_1, y_1, \ldots, x_{n+1}, y_{n+1})$ be the Cartesian coordinates on $\mathbb{R}^{2n+2}$ and set

$$\alpha_0 = \sum_{j=1}^{n+1}(x_j dy_j - y_j dx_j).$$

Then the restriction of α_0 to S^{2n+1} is a contact form, and we obtain a contact manifold $(S^{2n+1}, \xi_0 = \ker \alpha_0)$. This is called the *standard contact structure on* S^{2n+1}. The Reeb vector field of α_0 is given by

$$R_{\alpha_0} = \sum_{j=1}^{n+1} \left(x_j \frac{\partial}{\partial y_j} - y_j \frac{\partial}{\partial x_j} \right).$$

Regarding S^{2n+1} as the unit sphere in $\mathbb{C}^{n+1}$, ξ_0 coincides with its complex tangency. That is,

$$\xi_0 = TS^{2n+1} \cap J(TS^{2n+1}),$$

where J is the standard complex structure on $\mathbb{C}^{n+1}$. Moreover, the restriction of ξ_0 to the complement of one point $S^{2n+1} - \{p\}$ is contactomorphic to $(\mathbb{R}^{2n+1}, \xi_1)$ (see Proposition 2.1.8 in [53]).

By using the Moser trick or by other elementary methods, the following two fundamental facts can be proven.

Theorem 6.1 (Darboux Theorem) *Any contact $(2n + 1)$-manifold is locally contactomorphic to the standard contact structure $(\mathbb{R}^{2n+1}, \xi_1)$.*

By Theorem 6.1, for any point p on a contact manifold (M^{2n+1}, ξ), there exists a neighborhood U of p such that $(U, \xi|_U)$ is contactomorphic to $(\mathbb{R}^{2n+1}, \xi_1)$. Such a neighborhood is called a *Darboux chart*.

Theorem 6.2 (Gray Stability) *Let M be a closed manifold, and $\{\xi_t\}_{0 \le t \le 1}$ be a smooth family of contact structures on M. Then there exists an isotopy f_t of M such that $(f_t)_* \xi_0 = \xi_t$ for each $t \in [0, 1]$.*

The family $\{\xi_t\}_{0 \le t \le 1}$ in Theorem 6.2 is a smooth homotopy of hyperplane fields such that each ξ_t is a contact structure. Such a family is called a *smooth homotopy of contact structures*. Then Theorem 6.2 says that a smooth homotopy of contact structures on a closed manifold can be chased by a contact isotopy (an isotopy of contactomorphisms).

6.2 Contact 3-Manifolds

In this section, we give a brief overview of basic knowledge and important facts in 3-dimensional contact geometry. In what follows, $(M, \xi = \ker \alpha)$ always denotes a positive contact 3-manifold.

The Standard Contact 3-Sphere Let us review the most fundamental example, $(S^3, \xi_0 = \ker \alpha_0)$, from various viewpoints. First it can be seen as the strongly pseudoconvex boundary of a Stein domain $D^4 \subset \mathbb{C}^2$. In general, for a strictly plurisubharmonic function φ on a complex manifold, the real 1-form $i(\bar{\partial} - \partial)\varphi$ defines the canonical contact structure on each non-singular level set of φ, and its Reeb vector field is given by $J \left(\mathrm{grad}(\varphi) / \|\mathrm{grad}(\varphi)\|^2 \right)$. In our case, we can take the function

$$\varphi = \frac{1}{2}(|z_1|^2 + |z_2|^2) = \frac{1}{2}(x_1^2 + y_1^2 + x_2^2 + y_1^2).$$

Then we have indeed

$$i(\bar{\partial} - \partial)\varphi = \frac{i}{2}(z_1 d\bar{z}_1 + z_2 d\bar{z}_2 - \bar{z}_1 dz_1 - \bar{z}_2 dz_2) = -\mathrm{Im}(z_1 d\bar{z}_1 + z_2 d\bar{z}_2) = \alpha_0,$$

$$J\left(\text{grad}(\varphi)/\|\text{grad}(\varphi)\|^2\right)\big|_{\varphi=\frac{1}{2}} = x_1\frac{\partial}{\partial y_1} - y_1\frac{\partial}{\partial x_1} + x_2\frac{\partial}{\partial y_2} - y_2\frac{\partial}{\partial x_2} = R_{\alpha_0}.$$

We note that the integral curves of the Reeb vector field R_{α_0} are Hopf fibers. Hence, the standard contact structure ξ_0 is perpendicular to the Hopf fibration, and is invariant under the S^1-action $e^{i\theta}\cdot(z_1, z_2) = (e^{i\theta}z_1, e^{i\theta}z_2)$. These features can be interpreted in terms of Hamilton's quaternions. We regard $(z_1, z_2) = (x_1+iy_1, x_2+iy_2)$ as the quaternion $z_1 + z_2 j = x_1 + y_1 i + x_2 j + y_2 k$ and S^3 as the unit sphere of the quaternion ring $\mathbf{H} \cong \mathbb{R}^4$. Then it forms a Lie group for the multiplication of quaternions. The tangent space $T_1 S^3$ at $1 \in S^3$, which is the Lie algebra, is spanned by i, j and k. Then the contact plane at $1 \in S^3$ is spanned by j, k, and ξ_0 is invariant under the Lie right action. By the right action of S^3, the basis i, j, k of $T_1 S^3$ induces the three linearly independent right invariant vector fields

$$(-y_1, x_1, -y_2, x_2), \quad (-x_2, y_2, x_1, -y_1), \quad (-y_2, -x_2, y_1, x_1),$$

which give a tangential trivialization of S^3. Notice that the first vector field,

$$-y_1\frac{\partial}{\partial x_1} + x_1\frac{\partial}{\partial y_1} - y_2\frac{\partial}{\partial x_2} + x_2\frac{\partial}{\partial y_2},$$

coincides with the Reeb vector field R_{α_0}, which generates the Hopf fibration, and the other ones,

$$-x_2\frac{\partial}{\partial x_1} + y_2\frac{\partial}{\partial y_1} + x_1\frac{\partial}{\partial x_2} - y_1\frac{\partial}{\partial y_2}, \quad -y_2\frac{\partial}{\partial x_1} - x_2\frac{\partial}{\partial y_1} + y_1\frac{\partial}{\partial x_2} + x_1\frac{\partial}{\partial y_2},$$

span the standard contact structure ξ_0. Thus R_{α_0} and ξ_0 are indeed right invariant and described as $R_{\alpha_0} = i$ and $\xi_0 =< j, k >$, respectively. Moreover, the standard contact form is described as $\alpha_0 = i^*$, where i^*, j^*, k^* are the dual basis of i, j, k. Of course, we can also use the special unitary group $SU(2)$ for these descriptions through the following identification:

$$\rho\colon S^3 \to SU(2); \ (z_1, z_2) \mapsto \begin{pmatrix} z_1 & -\bar{z}_2 \\ z_2 & \bar{z}_1 \end{pmatrix}.$$

Here we have to notice that ρ inverts the order of products, namely, we have

$$\rho\big((z_1, z_2)\cdot(w_1, w_2)\big) = \rho(w_1, w_2)\rho(z_1, z_2).$$

Indeed, $(z_1, z_2)\cdot(w_1, w_2) = (z_1 w_1 - z_2\bar{w}_2, z_1 w_2 + z_2\bar{w}_1)$ corresponds to

$$\begin{pmatrix} w_1 & -\bar{w}_2 \\ w_2 & \bar{w}_1 \end{pmatrix}\begin{pmatrix} z_1 & -\bar{z}_2 \\ z_2 & \bar{z}_1 \end{pmatrix} = \begin{pmatrix} z_1 w_1 - z_2\bar{w}_2 & -\bar{z}_1\bar{w}_2 - \bar{z}_2 w_1 \\ z_1 w_2 + z_2\bar{w}_1 & \bar{z}_1\bar{w}_1 - \bar{z}_2 w_2 \end{pmatrix}.$$

Then the right action of $S^3 \subset \mathbf{H}$ corresponds to the left action of $SU(2)$. Therefore, when we regard S^3 as $SU(2)$, the contact structure ξ_0, the contact form α_0 and its Reeb vector field R_{α_0} become all left-invariant. This fact can be also explained in the following way. The matrices of the form $\begin{pmatrix} e^{i\theta} & 0 \\ 0 & e^{-i\theta} \end{pmatrix}$ form a subgroup of $SU(2)$ isomorphic to $U(1)$. Its right action on $SU(2)$ exactly coincides with the canonical S^1-action on S^3, and hence, the Hopf fibration is obtained as the natural projection $S^3 \cong SU(2) \to SU(2)/U(1) \cong \mathbb{C}P^1$. Therefore, the S^1-action and its generating vector field R_{α_0} are invariant under the left action of $SU(2)$ to itself. Moreover, the Hopf fibration and the standard contact structure ξ_0, which is perpendicular to the Hopf fibration, are also left-invariant.

Remark 6.1 Define $\mu \colon S^3 \to SU(2)$ by $\mu(z_1, z_2) = \begin{pmatrix} z_1 & z_2 \\ -\bar{z}_2 & \bar{z}_1 \end{pmatrix}$. Then it is a Lie group isomorphism, since we have

$$\mu\big((z_1, z_2) \cdot (w_1, w_2)\big) = \mu(z_1, z_2)\mu(w_1, w_2).$$

If we adopt μ instead of ρ as the identification, the right action of S^3 corresponds to the right action of $SU(2)$. However, this is not compatible with the usual convention that $SU(2)$ lineraly acts on $\mathbb{C}^2$ from the left. In this sense, it is natural to adopt the idenitification ρ in our case.

The Links of Simple Singularities As is well-known, all the finite subgroups of $SU(2)$ are classified. It contains all the finite cyclic groups and the binary triangle groups $\Gamma(2, 2, r), r \geq 2, \Gamma(2, 3, 3), \Gamma(2, 3, 4), \Gamma(2, 3, 5)$, and these are the only finite subgroups of $SU(2)$. This is proven through the classification of finite subgroups of $SO(3)$ and the double covering map $S^3 \cong SU(2) \to SO(3) \cong \mathbb{R}P^3$. Notice that $\Gamma(2, 2, r), \Gamma(2, 3, 3), \Gamma(2, 3, 4)$ and $\Gamma(2, 3, 5)$ are the double covers of the dihedral group D_r, the tetrahedral group T, the octahedral group O and the icosahedral group I, respectively. For each finite subgroup Γ, the contact structure ξ_0 descends to the compact quotient $\Gamma \backslash SU(2)$, since ξ_0 is invariant under the left action of $SU(2)$. In particular, we obtain the following two important examples.

Example 6.3 (Compact Quotients of the Standard Contact 3-Sphere)

1. In the case where Γ is the cyclic group C_n of order n generated by $\begin{pmatrix} e^{2\pi i/n} & 0 \\ 0 & e^{-2\pi i/n} \end{pmatrix} \in SU(2)$, the quotient $C_n \backslash SU(2)$ is the lens space $L(n, n-1)$ and ξ_0 descends to the canonical contact structure on it.
2. In the case where Γ is the so-called binary icosahedral group (order 120), namely, the binary triangle group $\Gamma(2, 3, 5)$, the quotient $\Gamma(2, 3, 5) \backslash SU(2)$ is the Poincaré homology sphere $M(2, 3, 5)$ and ξ_0 descends to the canonical contact structure on it.

Now we explain that such a contact manifold is the link of some hypersurface singularity $(V, \mathbf{0})$ in $\mathbb{C}^3$. Here the canonical contact structure on the link $L = V \cap S_\varepsilon^5$

is given by the complex tangency $TL \cap J(TL)$, which coincides with the standard contact structure (S_ε^5, ξ_0) restricted to L.

First, $SU(2)$ acts freely on $\mathbb{C}^2$ except at the only fixed point $\mathbf{0}$. Hence, each finite subgroup $\Gamma \subset SU(2)$ defines an isolated surface singularity $(\Gamma \backslash \mathbb{C}^2, \mathbf{0})$. Such a singularity is called a *simple singularity (Kleinian singularity or rational double point)*. Klein [94] investigated simple singularities from the viewpoint of polyhedral groups and invariant polynomials, and showed that they are analytically isomorphic to the following hypersurface singularities in $\mathbb{C}^3$:

1. A_k: $z_1^2 + z_2^2 + z_3^{k+1} = 0$ if $\Gamma = C_{k+1}$ $(k \geq 1)$,
2. D_k: $z_1^2 + z_2^2 z_3 + z_3^{k-1} = 0$ if $\Gamma = \Gamma(2, 2, k-2)$ $(k \geq 4)$,
3. E_6: $z_1^2 + z_2^3 + z_3^4 = 0$ if $\Gamma = \Gamma(2, 3, 3)$,
4. E_7: $z_1^2 + z_2^3 + z_2 z_3^3 = 0$ if $\Gamma = \Gamma(2, 3, 4)$,
5. E_8: $z_1^2 + z_2^3 + z_3^5 = 0$ if $\Gamma = \Gamma(2, 3, 5)$.

Then the quotient $\Gamma \backslash SU(2)$ is nothing but the link of each singularity. In particular, $C_n \backslash SU(2)$ and $\Gamma(2, 3, 5) \backslash SU(2)$ are the links of the A_{n-1}-singularity $z_1^2 + z_2^2 + z_3^n = 0$ and the E_8-singularity $z_1^2 + z_2^3 + z_3^5 = 0$, respectively.

Remark 6.2 (Various Singularity Links and 3-Dimensional Lie Groups) Milnor [126] extended Klein's work to the *Brieskorn singularities* $z_1^p + z_2^q + z_3^r = 0$. He showed that the 3-dimensional Brieskorn manifold $M(p, q, r)$, the link of $z_1^p + z_2^q + z_3^r = 0$, is a compact quotient of $SU(2)$, Nil^3 or $\widetilde{SL}(2; \mathbb{R})$ according as the rational number $p^{-1} + q^{-1} + r^{-1} - 1$ is positive, zero or negative. His work was later extended to the so-called *quasi-homogeneous singularities*, which are singularities with good $\mathbb{C}^*$-actions [32, 139, 149, 155, 159, 176]. Quasi-homogeneous singularities contain several important classes of singularities, such as simple singularities, Brieskorn singularities and also simple elliptic singularities. A *simple elliptic singularity* is a normal surface singularity such that the exceptional set of its minimal resolution consists of a single non-singular elliptic curve. Then its link is diffeomorphic to an S^1-bundle over the elliptic curve with negative Euler class, and hence, is a compact quotient of the Lie group Nil^3. By Saito [160], it is known that the following examples are the only hypersurface simple elliptic singularities:

1. $\tilde{E}_6$: $z_1^3 + z_2^3 + z_3^3 + \lambda_1 z_1 z_2 z_3 = 0$, $\lambda_1^3 + 27 \neq 0$,
2. $\tilde{E}_7$: $z_1^2 + z_2^4 + z_3^4 + \lambda_2 z_1 z_2 z_3 = 0$, $\lambda_2^4 - 64 \neq 0$,
3. $\tilde{E}_8$: $z_1^2 + z_2^3 + z_3^6 + \lambda_3 z_1 z_2 z_3 = 0$, $\lambda_3^6 - 432 \neq 0$.

On the other hand, the link of a cusp singularity, which is not quasi-homogeneous, is diffeomorphic to the quotient of the Lie group Sol^3 by a cocompact lattice, that is, a T^2-bundle over S^1 with hyperbolic monodromy (see [72, 107], see also [85, 132]). It is also proven by Karras [82] that the only hypersurface cusp singularities are given by the following form:

$$T_{pqr}: z_1^p + z_2^q + z_3^r + \lambda z_1 z_2 z_3 = 0, \quad p^{-1} + q^{-1} + r^{-1} < 1, \ \lambda \neq 0.$$

Neumann summarized such relations between 3-dimensional Lie groups and complex surface singularities. He also showed that such a singularity link carries the canonical CR structure induced by the left-invariant CR structure on the corresponding Lie group [33, 139]. In particular, the canonical contact structure on the link of a quasi-homogeneous singularity or a cusp singularity is the one induced by the left-invariant contact structure on the corresponding Lie group (for this topic, see also [86, 116, 162]).

Legendrian Knots and Transverse Knots

There are two important notions for knots in a contact 3-manifold.

Definition 6.3 (Legerndrian Knot, Transverse Knot) Let γ be a knot in M, namely an embedding $\gamma : S^1 \to M$. Then γ is called a *Legendrian knot* if $\gamma'(\theta) \in \xi$ for each $\theta \in S^1$. On the other hand, if $\alpha(\gamma'(\theta)) > 0$ (resp. < 0) for each $\theta \in S^1$, we call γ a *positive (resp. negative) transverse knot*.

Example 6.4 (Legendrian Knot) Let (x, y, z) be the coordinates on $\mathbb{R}^2 \times S^1$. Then $L = \{0\} \times S^1$ is a Legendrian knot for the contact structure defined by $\alpha = \cos z\, dx - \sin z\, dy$.

Example 6.5 (Transverse Knot) Let (r, θ, z) be the coordinates on $\mathbb{R}^2 \times S^1$, where (r, θ) are the polar coordinates on $\mathbb{R}^2$. Then $K = \{0\} \times S^1$ is a positive transverse knot for the contact structure defined by $\beta = dz + r^2 d\theta$.

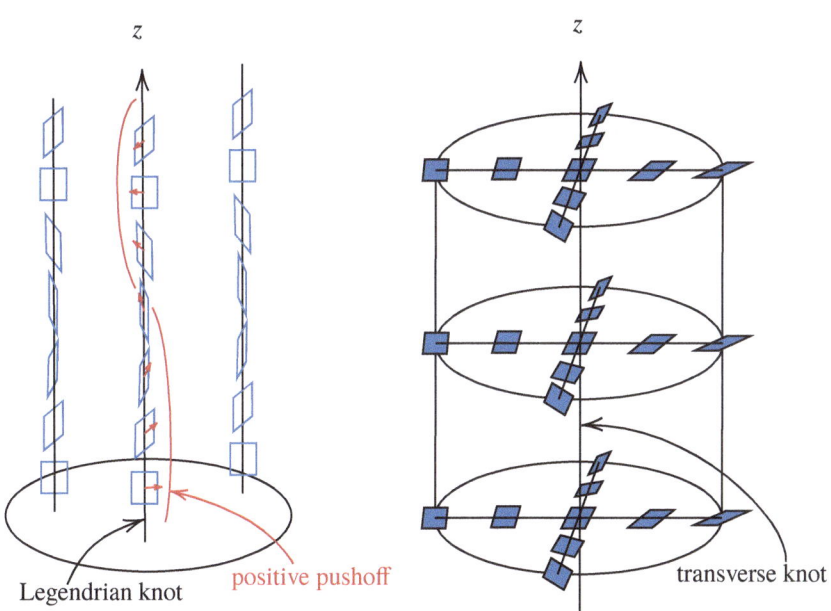

Legendrian knot | positive pushoff | transverse knot

In fact, Examples 6.4 and 6.5 give the standard neighborhoods of a Legendrian knot and a transverse knot, respectively.

Theorem 6.3 (Neighborhood Theorem for a Legendrian Knot) *Let γ be a Legendrian knot in (M, ξ). Then there exists a neighborhood of γ that is contactomorphic to some neighborhood of L in $(\mathbb{R}^2 \times S^1, \ker \alpha)$ of Example 6.4.*

Theorem 6.4 (Neighborhood Theorem for a Transverse Knot) *Let γ be a transverse knot in (M, ξ). Then there exists a neighborhood of γ that is contactomorphic to some neighborhood of K in $(\mathbb{R}^2 \times S^1, \ker \beta)$ of Example 6.5.*

Moreover, we can transform a given Legendrian knot to a transverse knot by an arbitrarily small smooth perturbation.

Definition 6.4 (Transverse Push-Off) By Theorem 6.3, a Legendrian knot in (M, ξ) can be identified with L in Example 6.4. Shifting the knot L in the direction of the vector field $X = \sin z \partial_x + \cos z \partial_y$, we obtain a positive transverse knot $\gamma_+(z) = (\delta \sin z, \delta \cos z, z)$ with $\delta > 0$. If we shift it in the opposite direction, we obtain a negative transverse knot $\gamma_-(z) = (-\delta \sin z, -\delta \cos z, z)$. We call these the *positive transverse push-off* and the *negative transverse push-off*, respectively.

Open Book Decomposition

Definition 6.5 (Open Book Decomposition) Let M be a closed orientable 3-manifold. Suppose there are a link K in M and a locally trivial fibration

$$p \colon M \setminus K \to S^1$$

such that there is a trivialized tubular neighborhood $\nu(K) \cong K \times D^2$ of K in which p coincides with the angular coordinate θ on $\nu(K) \setminus K$. Then (K, p) is called an *open book decomposition* of M. In this case, K is called the *binding* and the closure of each fiber $\overline{p^{-1}(\theta)}$ $(\theta \in S^1)$ is called the *page* (notice that a page is a compact surface with boundary K). Moreover, the monodromy of the fibration p is called the *monodromy* of the open book decomposition.

Alexander [3] showed that every closed orientable 3-manifold admits an open book decomposition. Combining with the following result, it proves the existence of a contact structure on any closed orientable 3-manifold.

Theorem 6.5 (Thurston and Winkelnkemper [174]) *Let M be closed oriented 3-manifold and (K, p) be an open book decomposition on it. Then there exists a contact form α on M such that K is a positive transverse knot of $\xi = \ker \alpha$ and $d\alpha$ is a positive volume form on each page.*

Remark 6.3 It was Martinet [117] who first proved the existence of a contact structure on any closed orientable 3-manifold. His proof was based on the theorem of Lickorish and Wallace and provided modification operations of a contact structure associated with Dehn surgeries. This idea later led to contact Dehn surgeries, whose formulation needs the convex surface theory (see Sects. 6.3 and 6.4).

An open book decomposition (K, p) of M and a contact structure $\xi = \ker \alpha$ are said to be *compatible* if they satisfy the conditions of Theorem 6.5. We also say that the open book decomposition *supports* the contact structure. In fact, the contact structure compatible with a given open book decomposition is unique up to contact isotopy, which is proven by using the Gray stability.

Example 6.6 The 3-sphere S^3 admits the trivial open book decomposition having D^2 as its page with trivial monodromy. The contact structure supported by this open book is the standard contact structure (S^3, ξ_0). Moreover, S^3 has another open book decomposition such that the binding is the Hopf link, the page is a positive Hopf band H_+, and the monodromy is the right-handed Dehn twist τ_c about the core c of H_+. Its compatible contact structure is again (S^3, ξ_0). These open book decompositions can be also seen as the Milnor fibrations associated with the polynomials z_1 and $z_1 z_2$ (or $z_1^2 + z_2^2$), respectively, where (z_1, z_2) is the coordinates on $\mathbb{C}^2$ and $S^3 \subset \mathbb{C}^2$ is regarded as the unit sphere.

Let S_1 and S_2 be compact oriented surfaces with boundaries. For each j ($j = 1, 2$), take a proper simple arc $\alpha_j \cong [-1, 1]$ in S_j with $\partial \alpha_j \subset \partial S_j$, and a tubular neighborhood $N(\alpha_j) \cong \alpha_j \times [-1, 1]$. Gluing S_1 and S_2 by identifying $N(\alpha_1)$ and $N(\alpha_2)$ through the orientation-preserving homeomorphism

$$h: \alpha_1 \times [-1, 1] \to \alpha_2 \times [-1, 1]; (s, t) \to (-t, s),$$

we obtain a new compact oriented surface S with boundary. This operation is called a *plumbing* of S_1 and S_2.

Example 6.7 Take two copies of Hopf bands H_1^+ and H_2^+. For each j ($j = 1, 2$), let c_j and α_j be the core curve and the cocore arc of H_j. Plumbing H_1 and H_2 for the arcs α_1 and α_2, we obtain a new compact surface S with boundary. Regarding S as an embedded surface in S^3, it is a once punctured torus bounded by the trefoil knot. In fact, S^3 admits an open book decomposition with the page S and the monodromy the composition of Dehn twists $\tau_{c_1} \circ \tau_{c_2}$. This can be seen as the Milnor fibration of the polynomial $z_1^2 + z_2^3$. Therefore, the compatible contact structure is again (S^3, ξ_0).

In general, for a given open book decomposition (K, p) of M, plumbing a positive Hopf band H^+ to its page $\overline{p^{-1}(\theta)}$ and composing its monodromy with the right-handed Dehn twist τ_c about the core c of H^+, we obtain a new open book decomposition of M. Such an operation is called a *positive stabilization*. This operation does not change the compatible contact structure, since it corresponds to the contact connected sum (see Definition 6.11) with the standard contact 3-sphere (S^3, ξ_0).

Giroux proved the converse of Theorem 6.5 in a complete form.

Theorem 6.6 (Giroux [58]) *Let (M, ξ) be a closed coorientable contact 3-manifold. Then there exists a unique supporting open book decomposition of (M, ξ) up to positive stabilization.*

Therefore, there exists the following one-to-one correspondence, which is called the *Giroux correspondence*. For the details, see [42, 53, 58, 151].

{contact structures on M}/contact isotopies

$\cong$ {open book decompositions of M}/positive stabilizations.

Tight vs. Overtwisted

The notions of tightness and overtwistedness play central roles in the 3-dimensional contact geometry. The following invariant for a Legendrian knot becomes important when the tightness of a contact structure is argued.

Definition 6.6 (Thurston–Bennequin Number) Let L be a homologically trivial Legendrian knot in (M, ξ), and L' be its shift in the normal direction of ξ. Then the linking number of L and L' is called the *Thurston–Bennequin number* of L, and is denoted by $TB(L)$. It can be also seen as the difference between the contact framing of L defined by ξ and the surface framing defined by a Seifert surface of L.

Now we are ready to define the important notions.

Definition 6.7 (Overtwisted Disk) An embedded disk D in (M, ξ) is called an *overtwisted disk* if it is bounded by a Legendrian knot whose Thurston–Bennequin number is 0, and the characteristic foliation $\xi \cap D$ has only one singular point inside D.

Definition 6.8 (Tight, Overtwisted) (M, ξ) is said to be *tight* if it does not contain an overtwisted disk. If it does, it is said to be *overtwisted*.

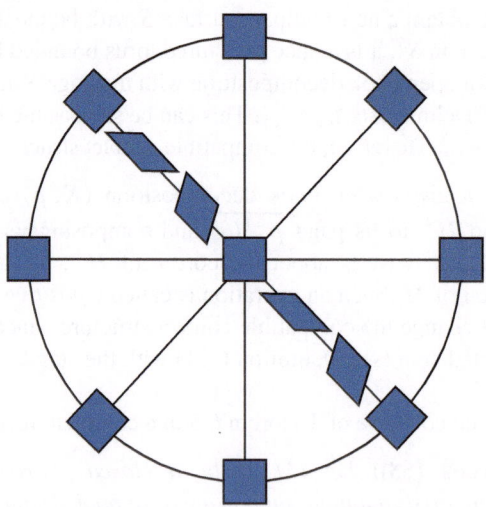

The standard contact 3-sphere (S^3, ξ_0) is tight (Bennequin [8]). In general, a contact manifold covered by a tight contact manifold is tight. Hence, for any finite subgroup $\Gamma \subset SU(2)$, the compact quotient $\Gamma \backslash SU(2)$ is tight. Thus the contact 3-

manifolds in Example 6.3 are tight (in fact, all the examples raised in Remark 6.2 are tight, since they are singularity links, and hence, Stein fillable, see Theorem 6.15). Moreover, removing one point from a tight contact manifold, the complement is still tight. Hence the standard contact structure on $\mathbb{R}^3$ is also tight. On the other hand, the following example contains an overtwisted disk.

Example 6.8 (Overtwisted Structure on $\mathbb{R}^3$) Let (r, θ, z) be the coordinates on $\mathbb{R}^3$, where (r, θ) is the polar coordinates on $\mathbb{R}^2$. We set $\alpha_{ot} = \cos r \, dz + r \sin r \, d\theta$. Then the disk $D_\pi \times \{*\}$ of radius π is an overtwisted disk in the contact structure $\xi_{ot} = \ker \alpha_{ot}$. Hence, $(\mathbb{R}^3, \xi_{ot})$ is overtwisted. In particular, it is not contactomorphic to the standard contact structure on $\mathbb{R}^3$.

Now we explain an operation modifying a given contact structure on a 3-manifold to a new overtwisted structure on the same manifold. Let (r, θ, z) be the coordinates on $\mathbb{R}^2 \times S^1$ as in Example 6.5, and take the contact form $\alpha = \cos r \, dz + r \sin r \, d\theta$ on $\mathbb{R}^2 \times S^1$. Then, just as in the above example, the solid torus $D_\pi \times S^1$ contains an S^1-family of overtwisted disks. This solid torus is called the *Lutz tube*.

Let K be a transverse knot in a closed contact 3-manifold (M, ξ). Then we can insert the Lutz tube into (M, ξ) along K as follows. First, by Theorem 6.4, there exists a tubular neighborhood $N(K)$ of K that is contactomorphic to $\left(D_R^2 \times S^1, \ker(dz + r^2 d\theta)\right)$ with R being a small positive number. We take a smaller solid torus $N_\varepsilon(K) = D_\varepsilon^2 \times S^1 \subset N(K)$ and remove it from M. Gluing $M \setminus N_\varepsilon(K)$ and the Lutz tube $D_\pi \times S^1$ along the boundary torus (in fact, we have to retake the tube so that the radius is a little larger than π), we obtain a new closed contact 3-manifold, which is obviously diffeomorphic to the original M and contains overtwisted disks. This operation is called the *Lutz twist*. If we use the tube $D_{2\pi} \times S^1$ instead of $D_\pi \times S^1$, we call the operation the *full Lutz twist*.

One might worry here whether the gluing is really possible so that the contact structures smoothly match up along the boundary. This is indeed possible, since the positive (resp. negative) contactness of a contact plane field is interpreted as the property that it rotates counterclockwise (resp. clockwise) when we go ahead along a Legendrian curve. Strictly speaking, we use the following proposition to smoothly glue the two pieces as contact manifolds.

Proposition 6.1 *Let $p, q \colon \mathbb{R}_{>0} \to \mathbb{R}$ be smooth functions. Then the 1-form $\alpha = p(r)dz + q(r)d\theta$ on $(\mathbb{R}^2 \setminus \{0\}) \times S^1$ defines a positive (resp. negative) contact structure if and only if the plane curve $c(t) = (p(t), q(t))$ rotates counterclockwise (resp. clockwise) around the origin in $\mathbb{R}^2$.*

Proof By a simple calculation, we obtain $\alpha \wedge d\alpha = \begin{vmatrix} p(r) & p'(r) \\ q(r) & q'(r) \end{vmatrix} dr \wedge d\theta \wedge dz$. Since the determinant in the right-hand side is the area velocity swept by the plane curve $c(t)$, it is positive (resp. negative) if and only if the curve rotates counterclockwise (resp. clockwise) around the origin. $\square$

Since $N(K) \setminus N_\varepsilon(K)$ is contactomorphic to $\left((D_R^2 \setminus D_\varepsilon^2) \times S^1, \ker(dz + r^2 d\theta) \right)$ and $-\alpha = -\cos r\, dz - r \sin r\, d\theta$ is also a contact form for the Lutz tube $D_\pi \times S^1$, we define functions $p(t)$ and $q(t)$ over $[0, \pi] \cup [\pi + \varepsilon, \pi + R]$ by

$$(p(t), q(t)) = \begin{cases} (-\cos t, -t \sin t) & (0 \le t \le \pi) \\ (1, (t - \pi)^2) & (\pi + \varepsilon \le t \le \pi + R). \end{cases}$$

Then it is clear that they can be smoothly extended over $[0, \pi + R]$ so that $(p(t), q(t))$ is in the first quadrant for each $\pi < t < \pi + \varepsilon$ and the plane curve $c(t) = (p(t), q(t))$ rotates counterclockwise around $\mathbf{0} \in \mathbb{R}^2$. Thus we can glue $M \setminus N_\varepsilon(K)$ and the Lutz tube $D_\pi \times S^1$ as contact manifolds. Though there is ambiguity for the choice of the plane curve $c(t)$, any such two curves are smoothly isotopic through an isotopy preserving the condition of the above proposition. Therefore, the resultant contact structure does not depend on the choice of $c(t)$ by the Gray stability.

Now let us see an application of Lutz twists and full Lutz twists.

Example 6.9 (Overtwisted Structures on S^3) Let $(z_1 = re^{i\theta_1}, z_2 = r_2 e^{i\theta_2})$ be the coordinates on $\mathbb{C}^2$ and we regard S^3 as the unit sphere in $\mathbb{C}^2$. Then the standard contact structure ξ_0 is described as the kernel of the 1-from $\alpha_0 = r_1^2 d\theta_1 + r_2^2 d\theta_2$. This contact plane field is rotationally symmetric and transverse to all the Hopf fibers. The flow lines generated by the vector field $r_2 \frac{\partial}{\partial r_1} - r_1 \frac{\partial}{\partial r_2}$ are the family of Legendrian arcs that connect the two Hopf fibers $L_1 = \{r_2 = 0\}$ and $L_2 = \{r_1 = 0\}$. Along each arc, the contact plane rotates in the anticlockwise direction by $\frac{\pi}{2}$. Now we apply the Lutz twist to (S^3, ξ_0) along the transverse knot L_1 to obtain a new contact structure ξ' on S^3. If we do instead the full Lutz twist along L_1, we obtain another contact structure ξ''. Then both ξ' and ξ'' are overtwisted.

On the other hand, they are not homotopic as plane fields. This is verified by the following argument. Since $S^3 (\subset \mathbb{H})$ is a Lie group, it is parallelizable. We take the tangential trivialization i, j, k invariant under the right action of S^3, and fix it. Then, by corresponding an oriented plane filed on S^3 to its normal vector field (if the field is given by a contact form, take its Reeb vector field, for example), a smooth map $S^3 \to S^2$ is uniquely determined up to homotopy. Therefore, we can identify the set of homotopy classes of oriented plane fields on S^3 with $[S^3, S^2] = \pi_3(S^2) \cong \mathbb{Z}$. It is known that the map corresponding a map $f: S^3 \to S^2$ to its Hopf invariant $H(f)$ gives the isomorphism $\pi_3(S^2) \cong \mathbb{Z}$. Here the Hopf invariant $H(f)$ is defined to be the linking number of the two links $\tilde{f}^{-1}(p_1)$ and $\tilde{f}^{-1}(p_2)$ in S^3, where $\tilde{f}$ is a generic smooth map homotopic to f, and p_j $(j = 1, 2)$ are its distinct regular values. In this sense, the homotopy class of an oriented plane field on S^3 can be described by the Hopf invariant. Since the standard contact structure ξ_0 is a plane field invariant under the right action of S^3, it is identified with a constant map, namely, $0 \in \pi_3(S^2) = \mathbb{Z}$. Moreover, it is not difficult to see that the Lutz twist along a Hopf fiber change the Hopf invariant by 1. Thus, ξ' is identified with $1 \in \pi_3(S^2) = \mathbb{Z}$. On the other hand, the full Lutz twist along a Hopf fiber does not change the

Hopf invariant. (In general, the full Lutz twist does not change the homotopy type of a plane field.) Therefore, ξ'' is homotopic to ξ_0 as an oriented plane field, while ξ' is not.

The supporting open book decomposition of (S^3, ξ') is the one corresponding to a negative Hopf band, namely, the open book such that the binding is a negative Hopf link, the page is a negative Hopf band H^-, and the monodromy is the left-handed Dehn twist τ_c^{-1} about the core c of H^-. From the viewpoint of confoliations, we note that the contact structure ξ' with reversed orientation on S^3 is homotopic as a plane field, through the Reeb foliation, to the positive standard contact structure (S^3, ξ_0) [133].

In fact, closed overtwisted contact 3-manifolds have been completely classified by the following result. For the definition and the classification of higher-dimensional overtwisted contact structures, see [14].

Theorem 6.7 (Eliashberg [34]) *Let M be a closed orientable 3-manifold. Then any oriented plane field on M is homotopic as a plane field to a unique overtwisted contact structure up to contact isotopy.*

Remark 6.4 Eliashberg's contribution to this theorem is the uniqueness part, which is the difficult part. On the other hand, the fact that any oriented plane field on M is homotopic to an overtwisted contact structure is usually called the Lutz–Martinet Theorem, since Martinet [117] proved the existence of a contact structure on every M, and based on it, Lutz showed the existence in every homotopy class of oriented plane fields by using Lutz twists. The outline of the argument is as follows. For the details, see [53, 151] and also [59]. First, as is well-known, a closed orientable 3-manifold is parallelizable. Hence, with a fixed trivialization of TM, the set of homotopy classes of oriented plane fields is identified with $[M, S^2]$. Then, the difference of homotopy classes of two oriented plane fields ξ and η are described by the so-called d_2-invariant and d_3-invariant through the obstruction theory. Namely, $d_2(\xi, \eta)$ is the obstruction to homotopy between ξ and η over the 2-skeleton $M \setminus \text{Int}(D^3)$, which lies in $H^2(M; \mathbb{Z})$, and if it vanishes, $d_3(\xi, \eta)$ appears as the secondary obstruction to extending the homotopy over D^3. In fact, however, the first obstruction can be eliminated by the Lutz twist along a transverse knot that is Poincaré dual to $d_2(\xi, \eta)$, and the second one by the Lutz twist along an appropriate transverse knot in a Darboux chart taken in D^3. Therefore, we can construct an overtwisted contact structure in every homotopy class of oriented plane fields on M.

On the other hand, the classification of tight contact 3-manifolds is incomplete. However, for tight contact structures, it is known that the classification can be reduced to that on each prime 3-manifold [25, 28]. This theorem is called the prime decomposition theorem. Moreover, on some prime 3-manifolds, tight contact structures are completely classified. The most famous result is the following.

Theorem 6.8 (Eliashberg [37]) *A tight contact structure on the 3-sphere is contactomorphic to the standard contact 3-sphere (S^3, ξ_0).*

The classification is completed also on $S^2 \times S^1$, T^3, the lens space, T^2-bundles over S^1, circle bundles over a closed orientable surface, and some small Seifert fibered 3-manifolds [37, 43, 54, 56, 57, 74, 75, 81]. In all these works, the so-called convex surface theory, which we overview in Sect. 6.3, plays a central role.

6.3 Convex Surface Theory

Let S be a smooth orientable surface in a positive contact 3-manifold $(M, \xi = \ker \alpha)$. The intersection with the contact plane ξ determines a singular 1-dimensional foliation ξ_S on S, called the *characteristic foliation* of S, whose singularities correspond to the points where the contact plane is tangent to S. On the other hand, there is a neighborhood of S diffeomorphic to $S \times (-1, 1)$, on which the contact form is described as $\alpha = \beta_t + g_t dt$, where t is the coordinate on $(-1, 1)$ and for each t, β_t and g_t are a 1-form and a function on S, respectively. In fact, ξ_S is characterized by the following property.

Proposition 6.2 *Let Ω be an area form on S and X be a vector filed on S such that $i_X \Omega = \beta_0$. Then the vector field X is tangent to the characteristic foliation ξ_S.*

Proof The characteristic foliation ξ_S is nothing but the kernel of β_0. On the other hand, we have $\beta_0(X) = i_X i_X \Omega = 0$. Hence, X is tangent to ξ_S.   □

The next theorem says that the isotopy type of the characteristic foliation ξ_S uniquely determines the contact structure on a neighborhood of a closed surface S.

Theorem 6.9 (Giroux [55]) *Let S be a closed surface in an orientable 3-manifold M and $\xi_0 = \ker \alpha_0$, $\xi_1 = \ker \alpha_1$ be two cooriented contact structures on M. If the isotopy type of the characteristic foliation ξ_S coincides for the two contact structures, then there are two neighborhoods $N_0(S)$ and $N_1(S)$ of S and a contactomorphism*

$$\varphi \colon (N_0(S), \xi_0|_{N_0(S)}) \to (N_1(S), \xi_1|_{N_1(S)})$$

such that φ is isotopic to $id_{N_0(S)}$ and $\varphi(S) = S$.

Proof First we take two neighborhoods of S diffeomorphic to $S \times (-1, 1)$ such that the two contact forms can be written as $\alpha_0 = \beta_t + g_t dt$ and $\alpha_1 = \gamma_t + h_t dt$. Then there is a diffeomorphism ψ between the two neighborhoods that preserves the height function and $\psi^* \gamma_0 = \beta_0$. Hence we may assume that the description $\alpha_0 = \beta_t + g_t dt$ and $\alpha_1 = \gamma_t + h_t dt$ can be obtained on the same neighborhood of S and $\beta_0 = \gamma_0$ originally. Now we put $\alpha_s = (1-s)\alpha_0 + s\alpha_1$ for $s \in [0, 1]$. Then there exists a small neighborhood of S on which α_s is a contact form for each s by the following argument. Since we have $\alpha_s = \beta_0 + ((1-s)g_0 + sh_0)dt$ along $S \times \{0\}$,

$$\alpha_s \wedge d\alpha_s$$
$$= \left(\beta_0 + (1-s)g_0 dt + sh_0 dt\right) \wedge \left(d\beta_0 + (1-s)(dg_0 - \dot\beta_0)\right)$$
$$\wedge dt + s(dh_0 - \dot\gamma_0) \wedge dt\big)$$
$$= (1-s)\left(g_0 d\beta_0 + \beta_0 \wedge (dg_0 - \dot\beta_0)\right) \wedge dt + s\left(h_0 d\beta_0 + \beta_0 \wedge (dh_0 - \dot\gamma_0)\right) \wedge dt$$
$$= (1-s)\alpha_0 \wedge d\alpha_0 + s\alpha_1 \wedge d\alpha_1 > 0$$

holds along $S \times \{0\}$. Therefore, α_s is a contact form on a small neighborhood of $S \times \{0\}$. Now the conclusion follows from the standard argument using the Moser trick. □

A vector field v on a positive contact 3-manifold $(M, \xi = \ker \alpha)$ is called a *contact vecor field* if it satisfies $\mathcal{L}_v\alpha = f\alpha$ for some smooth function $f: M \to \mathbb{R}$. This condition is equivalent to that the flow φ_t of v preserves the contact plane field ξ. Now we define the leading roles in the convex surface theory.

Definition 6.9 (Convex Surface, Dividing Set) A closed smooth surface S in (M, ξ) is said to be *convex* if there exists a contact vector field v transverse to S. The *dividing set* Γ_S of a convex surface S is defined by $\Gamma_S = \{x \in S \mid v(x) \in \xi_x\}$.

Example 6.10 We take the contact form $\alpha = dz + xdy - ydx$ on $\mathbb{R}^3$, which defines the standard contact structure. Let B_r be the 3-ball of radius r $(r > 0)$ centered at the origin $\mathbf{0} \in \mathbb{R}^3$. Then the boundary sphere $S_r = \partial B_r$ is a convex surface, since the contact vector field $v = x\partial_x + y\partial_y + 2z\partial_z$ is transverse to S_r. Indeed, we have

$$L_v\alpha = di_v\alpha + i_v d\alpha = d(2z + yx - xy) + 2i_v(dx \wedge dy)$$
$$= 2(dz + xdy - ydx) = 2\alpha,$$
$$v(x^2 + y^2 + z^2) = 2x^2 + 2y^2 + 4z^2 > 0.$$

Moreover, it is easily checked that the dividing set Γ coincides with the equator $\{z = 0\} \subset S_r$.

Example 6.11 Let L be a Legendrian knot in (M, ξ), and take its standard neighborhood $(D_\delta^2 \times S^1, \ker \alpha)$ as in Example 6.4. Then, the radial vector field $v = x\frac{\partial}{\partial x} + y\frac{\partial}{\partial y}$ is a contact vector field transverse to the boundary torus $T_\delta = \partial(D_\delta^2 \times S^1)$. Indeed, it satisfies

$$\mathcal{L}_v\alpha = di_v\alpha + i_v d\alpha = d(x \cos z - y \sin z) + x \sin z dz + y \cos z dz$$
$$= \cos z dx - \sin z dy = \alpha.$$

Hence, T_δ is a convex torus whose dividing set Γ consists of the two parallel curves γ_+ and γ_-, where $\gamma_\pm = \{(\pm\delta \sin z, \pm\delta \cos z, z) \mid z \in S^1\}$.

A vector field X on a closed orientable surface is said to be of *Morse–Smale type* if it satisfies the following conditions:

1. All the singularities and periodic orbits of X are non-degenerate.
2. The limit set of any point is either a singularity or a closed orbit.
3. There is no orbit connecting two saddle points.

By Peixoto's result [152, 153], a smooth vector field is generically of Morse–Smale type in the sense of C^∞-topology. On the other hand, Giroux [55] showed that a smooth surface is convex if the vector field associated with its characteristic foliation is Morse–Smale. Hence, any closed surface embedded in a contact 3-manifold can be made convex by a C^∞-small perturbation. The contact structure near a closed convex surface has the following nice description.

Proposition 6.3 *Let S be a closed convex surface in (M, ξ). Then S has a neighborhood diffeomorphic to $S \times (-1, 1)$ such that a contact form of ξ can be described as $\alpha = \beta + gdt$, where β and g are a 1-from and a function on S, respectively.*

Proof By assumption, there is a contact vector field v transverse to S. Hence, we can take the trivialization of a neighborhood of S using the flow φ_t of v. On this neighborhood $S \times (-1, 1)$, we have the description $\alpha = \beta_t + g_t dt$ of a contact form of ξ. By construction, the contact vector field is described as $v = \frac{\partial}{\partial t}$, and hence, we have

$$\mathcal{L}_v \alpha = f\alpha \iff \dot\beta_t + \dot g_t dt = f(\beta_t + g_t dt) \iff \dot\beta_t = f\beta_t, \ \dot g_t = fg_t$$

for some smooth function f. Now we set $F_t = \int_0^t f(x, u)du$ and take a new contact form $e^{-F_t}\alpha$ for ξ. Then we obtain

$$\mathcal{L}_v(e^{-F_t}\alpha) = \frac{\partial}{\partial t}(e^{-F_t}\beta_t) + \frac{\partial}{\partial t}(e^{-F_t}g_t)dt$$
$$= -fe^{-F_t}\beta_t + e^{-F_t}\dot\beta_t - fe^{-F_t}g_t dt + e^{-F_t}\dot g_t dt$$
$$= e^{-F_t}(\dot\beta_t - f\beta_t) + e^{-F_t}(\dot g_t - fg_t)dt = 0.$$

This completes the proof.   □

In the following, we always assume that S is a closed convex surface. The dividing set Γ_S of S is a collection of embedded circles transverse to the characteristic foliation, and it divides surface S into the two domains

$$S_+ = \{x \in S \mid \alpha(v(x)) > 0\} \quad \text{and} \quad S_- = \{x \in S \mid \alpha(v(x)) < 0\}.$$

Moreover, a vector field tangent to ξ_S expands a volume form on S_+, contracts it on S_- and points outward from S_+ along Γ_S. This is verified by the following argument. By the above proposition, we may assume that the contact form on $S \times (-1, 1)$ is described as $\alpha = \beta + g\,dt$ and the contact vector field is given by $v = \frac{\partial}{\partial t}$. Since

$$\alpha \wedge d\alpha = (\beta + g\,dt) \wedge (d\beta + dg \wedge dt) = dt \wedge (g\,d\beta + \beta \wedge dg)$$

is a volume form on $S \times (-1, 1)$, we can take an area form $\Omega_0 = g\,d\beta + \beta \wedge dg$ on S. Then we have $dg \neq 0$ along $\Gamma = g^{-1}(0)$, and so Γ is a 1-dimensional submanifold of S.

As in Proposition 6.2, we take a vector field X tangent to ξ_S given by $i_X \Omega_0 = \beta$. Notice that X vanishes exactly where β is singular. Moreover, we have $X(g) = -1$ along $\Gamma = g^{-1}(0)$, since the following holds:

$$\beta = i_X \Omega_0 = i_X(g\,d\beta + \beta \wedge dg) = g\,i_X d\beta + \beta(X)dg - X(g)\beta = -X(g)\beta.$$

In particular, X points outward from S_+ along Γ and Γ is transverse to the characteristic foliation ξ_S.

Now we take another area form $\frac{1}{|g|}\Omega_0$ defined on $S \setminus \Gamma = S_+ \cup S_-$. Then the vector field X expands it on S_+ and contracts it on S_-. Indeed, we have

$$\mathcal{L}_X \left(\frac{1}{g}\Omega_0 \right) = d \left(\frac{1}{g} i_X \Omega_0 \right) = d \left(\frac{\beta}{g} \right) = \frac{d\beta}{g} - \frac{dg}{g^2} \wedge \beta$$

$$= \frac{1}{g^2}(g\,d\beta + \beta \wedge dg) = \frac{1}{g^2}\Omega_0.$$

Though the area form $\frac{1}{|g|}\Omega_0$ never extends across Γ as it is, by modifying it appropriately near Γ, we can obtain an area form Ω defined on the whole S such that the vector field X expands Ω on S_+ and contracts it on S_- (see [55] and p. 231 of [53]). Based on these properties, we define the following notion.

Definition 6.10 Let $\mathcal{F}$ be a singular 1-dimensional foliation on S. A collection Γ of embedded circles in S is said to *divide* $\mathcal{F}$ if the following conditions are satisfied:

1. Γ is transverse to $\mathcal{F}$.
2. There are an area form Ω on S and a vector field v tangent to $\mathcal{F}$ such that $S - \Gamma = S_+ \sqcup S_-$ and v points outward from S_+ along Γ, where $S_+ = \{x \in S \mid \mathcal{L}_v \Omega > 0\}$ and $S_- = \{x \in S \mid \mathcal{L}_v \Omega < 0\}$.

In fact, in the convex surface theory, what really matters is the dividing set rather than the characteristic foliation itself. Giroux proved the following striking result, which is the core of his theory.

Theorem 6.10 (Giroux [55]) *Let S be a closed convex surface in (M, ξ) and ξ_S be its characteristic foliation. Suppose that $\mathcal{F}$ is another singular 1-dimensional* *foliation on S divided by* Γ_S. *Then there is a* C^0-*small isotopy* $\{f_t\}_{t\in[0,1]}$ *of embeddings* $S \to M$ *such that:*

1. f_0 *coincides with the original embedding.*
2. $f_t(S)$ *is a convex surface for each* $t \in [0, 1]$.
3. *The characteristic foliation of* $f_1(S)$ *coincides with* $f_1(\mathcal{F})$.

This theorem says that the dividing set determines the contact structure on a neighborhood of a convex surface, after a C^0-small isotopy of the surface. As an application of it, we can define the following operation.

Definition 6.11 (Contact Connected Sum) Let (M_1, ξ_1) and (M_2, ξ_2) be closed contact 3-manifolds. For each $j \in \{1, 2\}$, there exists a small 3-ball $D_j \subset M_j$ contactomorphic to B_r in Example 6.10 by the Darboux theorem. Let Γ_j be the dividing set on ∂D_j as described in Example 6.10. After possibly perturbing the boundary of D_j, we can glue $M_1 \setminus D_1$ and $M_2 \setminus D_2$ by a orientation-reversing diffeomorphism $f : \partial(M_1 \setminus D_1) \to \partial(M_2 \setminus D_2)$ that preserves the characteristic foliations by the above theorem. Thus we obtain a new closed contact 3-manifold whose underlying manifold is the connected sum $M_1 \# M_2$. This operation is called the *contact connected sum* of (M_1, ξ_1) and (M_2, ξ_2).

Advancing the viewpoint of Theorem 6.10, Giroux obtained the following important criterion for the tightness of a contact structure.

Theorem 6.11 (Giroux's Criterion) *Let S be a closed convex surface in* (M, ξ). *If* $S \neq S^2$, *S has a tight neighborhood if and only if* Γ_S *contains no contractible circle. If* $S = S^2$, *S has a tight neighborhood if and only if* Γ_S *is connected.*

This criterion has led to many classification results on tight contact structures. Among them, Kanda proved an important result (Theorem 6.12 below), which is needed for the notion of contact Dehn surgery. In order to state his theorem, we first define the slope of a loop on the 2-torus. A loop on the boundary of the solid torus $S^1 \times D^2$ is called the *meridian* if it bounds an embedded 2-disk in $S^1 \times D^2$. Strictly speaking, the meridian is uniquely determined only as a homology class in $H_1\left(\partial(S^1 \times D^2); \mathbb{Z}\right) = \mathbb{Z}^2$. We denote it by μ. In particular, the loop $\{*\} \times \partial D^2$ represents the meridian. Now we can take a loop on the torus that transversely intersects with the meridian loop at just one point. The homology class represented by such a loop is called the *longitude*, which we denote by λ. The longitude is not uniquely determined and there is an ambiguity of choice caused by Dehn twists about the meridian. Here we take a preferred longitude λ and fix it. Let c be a simple closed curve on the torus $\partial(S^1 \times D^2)$. Since μ and λ form a basis of $H_1\left(\partial(S^1 \times D^2); \mathbb{Z}\right) = \mathbb{Z}^2$, the homology class of c is written as $p\mu + q\lambda$ with coprime integers p and q. Then we call the ratio $p/q \in \mathbb{Q} \cup \{\infty\}$ the *slope* of c. Notice that the slope of the meridian and the longitude are ∞ and 0, respectively. Moreover, a simple closed curve of non-zero integer slope represents another longitude (different from λ). Now we are ready to state Kanda's result.

Theorem 6.12 (Kanda [81]) *Let p be an integer. Then a tight contact structure on the solid torus $S^1 \times D^2$ such that the boundary is a convex torus whose dividing set consists of two parallel curves of slope p is uniquely determined up to contact isotopy.*

Originally, Kanda obtained this result in his proof of the following classification of tight contact structures on T^3, which was obtained also by Giroux.

Theorem 6.13 (Kanda [81], Giroux [57]) *Any tight contact structure on the 3-torus is contactomorphic to (T^3, η_n) for some integer n, where η_n is defined by the contact form $\gamma_n = \cos(2n\pi z)dx - \sin(2n\pi z)dy$. Moreover, the two tight contact structures η_m and η_n are contactomorphic to each other if and only if $m = n$.*

Later Theorem 6.12 was generalized by Honda and Giroux independently into the classification of tight contact structures on the solid torus [57, 74].

6.4 Contact Dehn Surgery

Let K be a knot in a closed 3-manifold M with a given framing, and $N(K)$ be a tubular neighborhood of K in M. In this case, on the boundary $\partial N(K) \cong T^2$, the meridian μ is naturally determined and the longitude λ is also determined by the framing. Now we consider the following topological operation on M. We remove the solid torus $N(K) \cong S^1 \times D^2$ from M and glue back to $\overline{M - N(K)}$ by a diffeomorphism of the boundary 2-torus. Thus we obtain a new closed 3-manifold. Suppose that the meridian of the solid torus $N(K)$ is sent to $p\mu + q\lambda$ on $\partial(M - N(K))$ by the gluing diffeomorphism. Then it is not difficult to see that the diffeomorphism type of the resultant 3-manifold depends only on the rational number p/q (it depends on K and the framing, of course). Therefore, we may denote the new closed 3-manifold by $M_{p/q}(K)$. This operation is called the (p/q)-*Dehn surgery to M along K*.

We want to perform this operation in the contact category. Suppose that K is a Legendrian knot in (M, ξ). By Theorem 6.3, we can take a tubular neighborhood of K that is contactomorphic to $(N_\delta, \eta) = (D_\delta^2 \times S^1, \ker \alpha)$ in Example 6.4. On the torus boundary $T_\delta = \partial N_\delta$, we can take the meridian μ and the longitude λ determined by the contact framing of η. Moreover, T_δ is a convex torus whose dividing set consists of two parallel curves of slope 0.

Now we will perform the (p/q)-Dehn surgery to M along K for these μ, λ, and obtain a contact structure on the manifold $M_{p/q}(K)$ as follows. Since we already have the restriction of ξ on $M - N(K)$, we only need to extend it over the solid torus $S^1 \times D^2$ that is glued back. When $(p, q) = (1, q)$, the gluing diffeomorphism of the boundary torus is represented by $\begin{pmatrix} 1 & k \\ q & qk + 1 \end{pmatrix}$ for some integer k. Then the inverse diffeomorphism sends the longitude on $\partial(M - N(K))$ to $-k\mu + \lambda$ on $\partial(S^1 \times D^2)$. Thus the slope of the dividing curves on $\partial(S^1 \times D^2)$ is equal to $-$

k. Hence, by Theorem 6.12, the contact structure $\xi|_{M-N(K)}$ uniquely extends to a contact structure ξ' on the whole manifold $M_{1/q}(K)$ such that it is tight on the solid torus $S^1 \times D^2$. Therefore, the operation producing $(M_{1/q}(K), \xi')$ from (M, ξ) is well-defined. This is called the *contact $(1/q)$-surgery along K*. For this operation, Ding and Geiges proved the following interesting result.

Theorem 6.14 (Ding and Geiges [27]) *Any two closed connected positive contact 3-manifolds can be connected by a finite sequence of contact (± 1)-surgeries.*

This theorem will be the key in the proof of Theorem 7.1.

6.5 Symplectic Fillings

Here we will review various symplectic fillings. We note, however, that there are many important results that we do not mention, since our aim here is the preparation for Chap. 7.

Let (X, ω) be a symplectic $2n$-manifold, namely, ω is a non-degenerate closed 2-form on X. A vector field v on (X, ω) is called a *Liouville vector field* if $L_v\omega = \omega$. Suppose that X is compact with boundary. The boundary ∂X is called ω-convex (*resp. ω-concave*) if there exists a Liouville vector field v near ∂X, pointing outwards (resp. inwards) along ∂X. In this case, $\xi = \ker(i_v\omega|_{\partial X})$ defines a positive (resp. negative) contact structure on the boundary ∂X.

Definition 6.12 Let (M, ξ) be a closed connected positive contact 3-manifold.

1. A *Stein filling* of (M, ξ) is a compact Stein domain W whose contact boundary ∂W is contactomorphic to (M, ξ). In this case we say that (M, ξ) is *Stein fillable*.
2. A *strong symplectic filling* of (M, ξ) is a compact symplectic manifold (X, ω) whose boundary is ω-convex and contactomorphic to (M, ξ). In this case we say that (M, ξ) is *strongly symplectically fillable*.
3. A *weak symplectic filling* (*symplectic filling*) of (M, ξ) is a compact symplectic manifold (X, ω) such that $\partial X = M$ as oriented manifolds and $(\omega_{|\xi})^{n-1} > 0$. In this case we say that (M, ξ) is *weakly symplectically fillable*, or simply, *symplectically fillable*.

The following implications are immediate by definition.

Stein fillable $\Rightarrow$ strongly symplectically fillable $\Rightarrow$ weakly symplectically fillable.

It is known that these implications are not invertible in general. Moreover, Eliashberg and Gromov have proven the following important result, which implies that the symplectic fillability is a strong condition for a contact 3-manifold.

Theorem 6.15 (Eliashberg and Gromov [38]) *If a closed positive contact 3-manifold is weakly symplectically fillable, then it is tight.*

Among these types of symplectic fillings, Stein fillings are most extensively studied and there are many known results. For Stein fillable contact 3-manifolds, the following characterization is known.

Theorem 6.16 (Loi and Piergallini [115]) *A contact 3-manifold (M, ξ) is Stein fillable if and only if it is supported by an open book decomposition whose monodromy is a composition of right-handed Dehn twists.*

Let (M, ξ) be a contact 3-manifold obtained from a Stein fillable contact 3-manifold by operating only contact (-1)-surgeries. Then (M, ξ) is also Stein fillable, since the 2-handle attaching in Eliashberg–Gompf's construction corresponds to the contact (-1)-surgery along the attaching Legendrian knot. Through these characterizations, Stein fillable contact 3-manifolds have been investigated from topology, geometry and analysis.

On the other hand, the classification problem of Stein fillings of a fixed Stein fillable contact 3-manifold have been also studied. The major known results are as follows.

1. Any Stein filling of (S^3, ξ_0) is diffeomorphic to D^4 (Eliashberg [36]).
2. The classification of Stein fillings of the standard contact structure on a lens space $L(p, q)$ was given by Lisca [113]. Recently, the complete description of the diffeomorphism type of minimal strong symplectic fillings of any tight contact structure on $L(p, q)$ has been given by Etnyre and Roy [45] and Christian and Li [23].
3. Any Stein filling of the Stein fillable contact structure on T^3 is diffeomorphic to $T^2 \times D^2$ (Stipsicz [168]). All the strong symplectic fillings of T^3 are equivalent up to symplectic deformation and blow-up (Wendl [177]).
4. Any Stein filling of the canonical contact structure of a simple singularity link is diffeomorphic to the Milnor fiber (Ohta and Ono [145, 147]). Moreover, a minimal strong symplectic filling of the link of a simple elliptic singularity is diffeomorphic to the Milnor fiber or the minimal resolution if the degree k satisfies $1 \le k \le 9$, and to the minimal resolution if $k \ge 10$ (Ohta and Ono [146]). (Combining with Theorem 5.10, the latter half can be also seen as the classification of Stein fillings.)
5. There are infinitely many contact 3-manifolds, each of which admits infinitely many pairwise homotopically inequivalent Stein fillings (Ozbagci and Stipsicz [150]). There are also infinitely many contact 3-manifolds, each of which admits infinitely many pairwise homeomorphic but non-diffeomorphic simply-connected Stein fillings (Akhmedov, Etnyre, Mark and Smith [2]). Moreover, there is an infinite family of contact 3-manifolds, each member of which admits a Stein filling with arbitrarily large Euler characteristic and arbitrarily small signature (Baykur and Van Horn Morris [7]).

For Stein fillings of higher-dimensional contact manifolds, there are also some known results. For example, Eliashberg–Floer–McDuff's result [121] states that under an aspherical assumption, any strong symplectic filling of the standard contact sphere is diffeomorphic to the ball. In this direction, there are recent developments by Barth, Geiges and Zehmisch [6] and by Lazarev [108]. On the other hand, there is also a non-uniqueness result by Oba [142], which shows that for any $n \geq 2$, there is an infinite family of contact $(4n - 1)$-manifolds, each member of which admits infinitely many pairwise homotopically inequivalent Stein fillings.

6.6 Symplectic Cobordisms and Concave Symplectic Fillings

Now that we have reviewed symplectic fillings, let us also look at symplectic cobordisms.

Definition 6.13 Let (M_1, ξ_1) and (M_2, ξ_2) be closed positive contact 3-manifolds.

1. A *complex cobordism* from (M_1, ξ_1) to (M_2, ξ_2) is a compact complex surface X such that the boundary ∂X consists of a strongly pseudoconvex part $\partial_+ X$ and a strongly pseudoconcave part $\partial_- X$ which are contactomorphic to (M_2, ξ_2) and $(-M_1, \xi_1)$, respectively. In this case we say that (M_1, ξ_1) is *complex cobordant* to (M_2, ξ_2).
2. A *symplectic cobordism* from (M_1, ξ_1) to (M_2, ξ_2) is a compact symplectic 4-manifold X such that the boundary ∂X consists of a ω-convex part $\partial_+ X$ and a ω-concave part $\partial_- X$ which are contactomorphic to (M_2, ξ_2) and $(-M_1, \xi_1)$, respectively. In this case we say that (M_1, ξ_1) is *symplectically cobordant* to (M_2, ξ_2).
3. A *Stein cobordism* from (M_1, ξ_1) to (M_2, ξ_2) is a complex cobordism from (M_1, ξ_1) to (M_2, ξ_2) admitting a strictly plurisubharmonic function φ such that M_1 and M_2 are non-singular level sets of φ. In this case we say that (M_1, ξ_1) is *Stein cobordant* to (M_2, ξ_2).

Notice that a Stein cobordism is a symplectic cobordism, since there are a symplectic form $\omega = i\partial\bar{\partial}\varphi$ and a Liouville vector field $v = \mathrm{grad}(\varphi)$ transverse to the boundary.

A strong symplectic filling can be glued together with a symplectic cobordism to produce a new symplectic filling, if the corresponding boundaries are contactomorphic via an orientation-reversing diffeomorphism. This is based on the following proposition, which is proven by an easy argument using the Moser trick.

Proposition 6.4 *Let S be a closed hypersurface in a symplectic manifold (X, ω) which is transverse to a Liouville vector field v (such a hypersurface is said to be of contact type). Then, S has a small neighborhood that is symplectomorphic to a piece of the symplectization of the contact manifold $(S, \ker(i_v\omega|_S))$.*

On the other hand, a weak symplectic filling cannot be necessarily glued with a symplectic cobordism, since there is no guarantee of the existence of a Liouvile vector field that is transverse to the boundary. Now let us define another type of symplectic filling.

Definition 6.14 Let (M, ξ) be a closed positive contact 3-manifold. A *concave symplectic filling* of (M, ξ) is a compact symplectic manifold (X, ω) whose boundary is ω-concave and contactomorphic to $(-M, \xi)$. In this case, (M, ξ) is said to be *concavely symplectically fillable*.

With respect to Proposition 6.4, a concave symplectic filling is also a good match with a symplectic cobordism. The simplest example of a concave symplectic filling is given as follows.

Example 6.12 Let (X, ω) be a closed symplectic 4-manifold. Fix a point $p \in X$ and take a small convex 4-ball B containing p. Then, $X - \text{Int} B$ is a concave symplectic filling of the standard contact 3-sphere.

A more interesting example is given by the following result, which is an application of Lempert's result (Theorem 5.9).

Theorem 6.17 (Lisca and Matić [114]) *Let V be a Stein domain with boundary. Then there exists a Kähler embedding of $\text{Int} V$ into a projective manifold of the same complex dimension.*

Hence, a contact 3-manifold admits a concave symplectic filling if it is Stein fillable. Indeed, if V is a Stein filling of a contact 3-manifold (N, η), the complement of an embedded image of V in a projective surface is a concave symplectic filling of (N, η). Using this result, Etnyre and Honda showed the following theorem.

Theorem 6.18 (Etnyre and Honda [44]) *Any closed positive contact 3-manifold admits infinitely many concave symplectic fillings.*

This means that concave symplectic fillings are not restrictive at all, which is in contrast to the convex case (Theorem 6.15). Their strategy is as follows. They first proved the next proposition, based on Eliashberg's method of convex handle attachment.

Proposition 6.5 *Any closed positive contact 3-manifold is Stein cobordant to a Stein fillable contact 3-manifold.*

Now let (M, ξ) be a closed positive contact 3-manifold, and take a Stein cobordism X from (M, ξ) to a Stein fillable contact 3-manifold (N, η). Then, X can be glued together with a concave symplectic filling W of (N, η) to obtain a concave symplectic filling of (M, ξ).

We note that Theorem 6.18 was later reproven by Ding–Geiges. They proved Proposition 6.5 based on the fact that the attachment of a Weinstein 2-handle along the convex (resp. concave) boundary corresponds to a contact (-1)-surgery (resp. $(+1)$-surgery). However, the subsequent discussion is the same as that of Etnyre–Honda. After all, the key point is that thanks to Proposition 6.4, the gluing of symplectic cobordisms is relatively easy.

On the other hand, it is not easy to glue a complex cobordism to a strongly pseudoconvex surface, since the CR structures on the corresponding boundaries do not necessarily coincide even if they are contactomorphic. In other words, the gluing of complex structures is far more difficult than that of symplectic structures. This difference is very important when we compare Kasuya–Zuddas' result with Etnyre–Honda's.

Chapter 7
Strongly Pseudoconcave Surfaces and Their Boundaries

In this chapter, we explain our result on the contact boundary of a strongly pseudoconcave surface (Theorem 7.1).

7.1 Concave Holomorphic Fillings of Contact 3-Manifolds

First, let us recall the definition of strongly pseudoconvex and pseudoconcave surfaces.

Definition 7.1 (Strongly Pseudoconvex Surface, Strongly Pseudoconcave Surface) Let (V, J) be a compact complex surface with smooth boundary. If there exists a collar neighborhood U of ∂V and a strictly plurisubharmonic function $\varphi \colon U \to \mathbb{R}$ without critical points such that

$$\partial V = \varphi^{-1}(0) \text{ and } U = \{\varphi \leq 0\}, \ \big(U = \{\varphi \geq 0\}\big),$$

then we say that V is a *strongly pseudoconvex (resp. pseudoconcave) surface*, and ∂V is the *strongly pseudoconvex (resp. pseudoconcave) boundary*. Moreover, φ is called a *defining function* of V.

We note that V is strongly pseudoconvex (resp. pseudoconcave) if and only if the complex tangency of the boundary ∂V, namely $\xi = T\partial V \cap J(T\partial V)$, is a positive (resp. negative) contact structure on the 3-manifold ∂V. We are interested in the contact 3-manifolds that arise as the boundary of such complex surfaces.

In what follows, (M, ξ) denotes a positive closed contact 3-manifold.

Definition 7.2 (Holomorphic Filling, Concave Holomorphic Filling) A *holomorphic filling* of (M, ξ) is a strongly pseudoconvex surface whose boundary is contactomorphic to (M, ξ). Similarly, a *concave holomorphic filling* of (M, ξ) is a strongly pseudoconcave surface whose boundary is contactomorphic to $(-M, \xi)$.

Now let us consider the following problem.

Problem 7.1 Which positive closed contact 3-manifold admits a holomorphic filling?

This problem has already been solved by Bogomolov and de Oliveira (Theorem 5.10). Indeed, they showed that a holomorphically fillable CR 3-manifold is deformation equivalent to a Stein fillable one. By the Gray stability, the underlying contact structure on the boundary does not change through the deformation. Then it follows that a holomorphically filable contact 3-manifold is Stein fillable. As a consequence, the holomorphic fillability and the Stein fillability are equivalent for contact 3-manifolds. Therefore contact structures that can be realized as the boundary of a strongly pseudoconvex surface are very restrictive.

Then it is natural to ask about the case of strongly pseudoconcave surfaces.

Problem 7.2 Which positive closed contact 3-manifold admits a concave holomorphic filling?

By Theorem 5.9, it is obvious that a Stein fillable contact manifold admits a concave holomorphic filling. Hence, we have to consider non-Stein fillable contact 3-manifolds. In particular, it is of interest whether an overtwisted contact 3-manifold admits a concave holomorphic filling. In this sense, the following is an interesting example.

Example 7.1 (Kasuya and Zuddas [87]) There is a collar neighborhood C of the boundary of $E(\rho_1, \rho_2)$ such that $E(\rho_1, \rho_2) - C$ is a non-Kähler strongly pseudoconcave surface. Moreover, the contact structure on the boundary 3-sphere is the overtwisted structure ξ' in Example 6.9 with the reversed orientation.

This example should seem natural to those who well understood the Giroux correspondence and Di Scala–Kasuya–Zuddas's construction. For, the open book decomposition of S^3 corresponding to the negative Hopf band naturally arises on the boundary of the standard neighborhood of a negative Lefschetz singularity. Indeed, the contact structure on the strongly pseudoconcave boundary is the overtwisted structure supported by this open book. As far as the author knows, this is the first example of a concave holomorphic filling of an overtwisted contact 3-manifold. Also the existence of such an example was the original motivation for us to work on Problem 7.2. Now let us show the complete answer to Problem 7.2, which is the following theorem.

Theorem 7.1 (Kasuya and Zuddas [88]) *Any positive closed contact 3-manifold admits infinitely many concave holomorphic fillings.*

This means that there is no obstruction for contact structures to admit concave holomorphic fillings, which is in contrast to the case of strongly pseudoconvex surfaces. This result is similar to those of Kiremidjian and Epstein–Henkin in the sense that it points out the "flexibility" of strongly pseudoconcave boundaries. Moreover, it can be also considered as the complex version of Etnyre–Honda's result.

7.2 Proof of Theorem 7.1

The outline of the proof is as follows. Based on Eliashberg's construction of Stein manifolds, we establish the method of holomorphic handle attaching to the strongly pseudoconcave boundary of a complex surface. The two methods are similar in many aspects as a matter of course, however, there is an important difference. In Eliashberg's method, a convex handle whose core is a totally real disk is attached to the strongly pseudoconvex boundary along a Legendrian knot, while in our case, we attach a concave handle whose core is a holomorphic disk to the strongly pseudoconcave boundary along a transverse knot. Thanks to the transversality of the attaching circle, in contrast to Eliashberg's case, our handle attaching can realize both contact (± 1)-surgeries as the change of the contact structure on the concave boundary. Then we can apply Ding–Geiges' theorem (Theorem 6.14), and realize any (M, ξ) as the boundary of a strongly pseudoconcave surface by an iteration of our concave handle attaching. Now we explain the detail following the argument in [88].

Concave Holomorphic Handle
In order to give the model of a concave holomorphic 2-handle, we make use of the following well-known fact.

Lemma 7.1 *Let $\Omega \subset \mathbb{C}^2$ be the domain defined by*

$$\Omega = \left\{ (z_1, z_2) \in \mathbb{C}^2 \mid |z_2| < \exp(\varphi(z_1)) \right\}.$$

Then, the boundary $\partial \Omega$ is strongly pseudoconvex (resp. pseudoconcave) if and only if the function $-\varphi$ (resp. φ) is strictly subharmonic.

Example 7.2 (Concave Holomorphic Handle) Let

$$\Omega_a = \left\{ (z_1, z_2) \in \mathbb{C}^2 \mid |z_2| \le \exp\left(\frac{|z_1|^2}{a} - a \right) \right\}$$

for any $a > 0$. Then, the boundary $\partial \Omega_a$ is strongly pseudoconcave. Hence, setting $H_a = \Omega_a \cap \{ |z_1| \le 1 + a^{-1} \}$, this is a concave holomorphic handle.

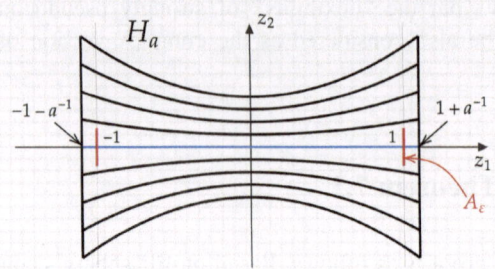

Since the domain Ω_a is foliated by strongly pseudoconvex surfaces $\{\partial\Omega_c \mid c \geq a\}$ and the z_1-axis, there exists a strictly plurisubharmonic function ψ on $\Omega_a - \{z_2 = 0\}$ such that $\psi^{-1}(0) = \partial\Omega_a$ and $\psi^{-1}([0, \infty)) = \Omega_a - \{z_2 = 0\}$. With a slight abuse of notation, we also denote by ψ the function restricted to $H_a - \{z_2 = 0\}$.

Handle Attaching to the Strongly Pseudoconcave Boundary

Let W be a strongly pseudoconcave surface with smooth real analytic boundary, ζ the contact structure on the boundary ∂W, and L a Legendrian knot in $(\partial W, \zeta)$. Moreover, we take a (strictly plurisubharmonic) defining function φ of W. Now we attach the holomorphic handle H_a to W in the following manner.

First, we take a positive transverse push-off K of L in $(\partial W, \zeta)$, and take the standard neighborhood $(D^2 \times S^1, \ker \beta)$ described in Example 6.5 (Theorem 6.4), and then, a sufficiently thin annulus in the solid torus whose core is K is a totally real annulus in W, since it is transverse to ζ. Now, we take a sufficiently thin annulus A_ε in H_a such that

$$A_\varepsilon = S^1 \times (-\varepsilon, \varepsilon) \subset H_a \subset \mathbb{C} \times \mathbb{C},$$

and define $g_n \colon A_\varepsilon \to D^2 \times S^1 \subset \mathbb{C} \times S^1 \subset \mathbb{C}^2$ by

$$g_n(z_1, z_2) = (z_1^n z_2, z_1).$$

Then, g_n is a diffeomorphism between the two totally real annuli.

By Whitney's theorem [178], we can perturb g_n to a real analytic embedding

$$f_n \colon A_\varepsilon \to D^2 \times S^1 \subset W$$

by a C^∞ approximation. Since a real analytic diffeomorphism between totally real submanifolds extends to their sufficiently small neighborhoods as a biholomorphism, there exists a biholomorphism $\widetilde{f}_n \colon U \to V$ that is an extension of f_n, where U and V are some neighborhoods of $A_\varepsilon \subseteq \mathbb{C}^2$ and $f_n(A_\varepsilon) \subset W$, respectively. Now we can attach the handle H_a to W by $\widetilde{f}_n$ for a sufficiently large a such that the attaching region of H_a is contained in U. Namely, we obtain a new complex surface W' by $W' = W \cup_{\tilde{f_n}} H_a$.

Remark 7.1 Strictly speaking, the two annuli A_ε and $f_n(A_\varepsilon)$ must be in the interior of H_a and W, respectively, so that the gluing regions U and V can be taken as tubular neighborhoods of them. The annulus A_ε is surely included in $\text{Int}(H_a)$, since the unit circle S^1 is inside $\{|z_1| < 1 + a^{-1}\}$. On the other hand, the embedded annulus $f_n(A_\varepsilon)$ is on the boundary ∂W as it is. In order to have $f_n(A_\varepsilon)$ inside W, we need to retake the original Legendrian knot L inside W from the beginning. Concretely, we take a non-singular level set $\varphi^{-1}(d)$ of the defining function φ with small $d > 0$. Then it is inside W and contactomorphic to the original contact boundary ∂W, so we can take the Legendrian knot L on it. We omitted such an argument so that the notations do not become too heavy.

At this stage, the complex surface W' has corners along a 2-torus in the boundary, corresponding to the transverse intersection $\partial W \cap \partial H_a$. Next, we need to remove the corners to make the boundary smooth.

Making the Boundary Smooth

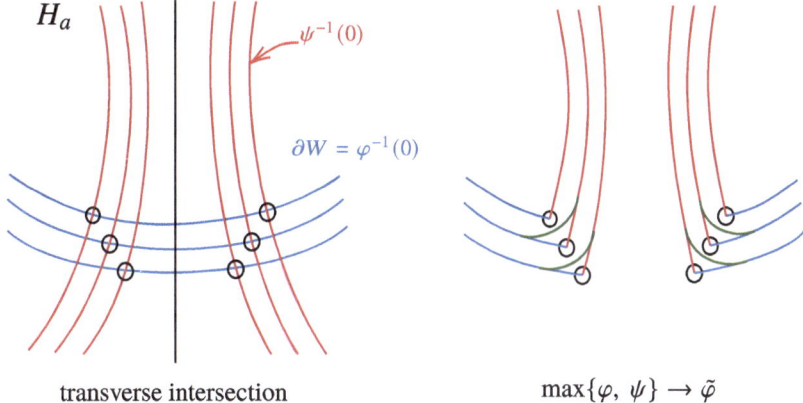

transverse intersection $\max\{\varphi, \psi\} \rightarrow \tilde{\varphi}$

Recall that φ is the defining function of W and ψ is the strictly plurisubahrmonic function on $H_a - \{z_2 = 0\}$. Since the corner of the boundary of W' is nothing but the transversal intersection of ∂W and ∂H_a, $\partial W'$ is the zero level set of the function $\max\{\varphi, \psi\}$. Since φ and ψ are both strictly plurisubharmonic, $\max\{\varphi, \psi\}$ is a continuous strictly plurisubharmonic function. By Richberg's theorem (Theorem 5.6), we can approximate it from above by a smooth strictly plurisubharmonic function $\tilde{\varphi}$. Since the level set $\tilde{\varphi}^{-1}(d)$ with small $d > 0$ is a smooth strongly pseudoconcave hypersurface in W', we can take a strongly pseudoconcave surface $\tilde{W}$ with smooth boundary such that $\tilde{W} \subset \text{Int}(W')$ and $\partial \tilde{W} = \tilde{\varphi}^{-1}(d)$.

Finally, we show that the contact structure η on $\partial \tilde{W}$ is contactomorphic to the resultant contact structure obtained from $(\partial W, \zeta)$ by a contact Dehn surgery along L. Topologically, the change from ∂W to $\partial \tilde{W}$ associated with the handle attaching of H_a corresponds to some p-Dehn surgery along L. Though the handle is attached along the transverse knot $f_n(S^1 \times \{0\})$, which is a perturbation of the positive push-off K of L, all the operations can be done in the standard neighborhood $(N_\delta, \ker \alpha)$ of the Legendrian knot L so that the change of the contact structure associated with the attaching of H_a occurs only inside $(N_\delta, \ker \alpha)$. Since the solid torus $\partial \widetilde{W} \cap H_a$ is a strongly pseudoconvex hypersurface in $H_a \subset \mathbb{C}^2$, the contact structure on it is tight. Therefore, $(\partial \widetilde{W}, \eta)$ is indeed the resultant contact manifold obtained from $(\partial W, \zeta)$ by a contact Dehn surgery along L. Since the longitude of the Dehn surgery is determined by the integer n of the embedding f_n, contact (± 1)-surgeries can be realized by retaking $n \in \mathbb{Z}$ if necessary. Thus both contact (± 1)-surgeries are realized by holomorphic handle attaching of H_a.

Since any (M, ξ) can be obtained from the standard contact 3-sphere (S^3, ξ_0) by a finite iteration of contact (± 1)-surgeries (Theorem 6.14), we obtain a concave holomorphic filling of (M, ξ) by realizing the sequence of surgeries by holomorphic handle attaching to a concave holomorphic filling of (S^3, ξ_0). This completes the proof of Theorem 7.1.

7.3 Kähler Metrics on Concave Holomorphic Fillings

Here we show that in Theorem 7.1 there exist both Kähler and non-Kähler fillings for any contact structure. It is clear that there exist non-Kähler fillings. For, if the original surface W is non-Kähler, then $\widetilde{W}$ is also non-Kähler. As an example of a non-Kähler W, we can take the complement of a small strongly pseuodoconvex 4-ball in a compact non-Kähler complex surface. Also we may use Example 7.1 as W. In this case, the boundary $(\partial W, \zeta)$ is not the standard contact 3-sphere, but there is no problem, since Theorem 6.14 gives a sequence of contact (± 1)-surgeries between arbitrarily two closed connected coorientable contact 3-manifolds.

On the other hand, we need the following lemma to make the surface $\widetilde{W}$ Kähler.

Lemma 7.2 *Let W be a Kähler complex surface with strongly pseudoconcave boundary, and $\widetilde{W}$ the strongly pseudoconcave surface obtained from W by attaching a holomorphic 2-handle H_a. Then, for a sufficiently large positive number a, the Kähler form on W can be extended over H_a.*

Proof Let ω be the Kähler form on W. Since the attaching circle of H_a has a Stein tubular neighborhood U in $\widetilde{W}$, we obtain an extension of the Kähler form over U, which we still denote by ω, admitting a Kähler potential h defined on U. Namely, we have $\omega = i\partial\bar{\partial}h$ on U. By retaking a large enough, we may assume that the attaching region of H_a is contained in U and h is defined over a thin region $H_a \cap \{ 1 - 3\varepsilon \leq |z_1| \leq 1 \}$ for some small $\varepsilon > 0$. Now we set

$$h_1(z_1, z_2) = h(z_1, z_2) + b \log |z_1| + c$$

for positive numbers b and c. Then it is a strictly plurisubharmonic function which is defined on the thin region and satisfies $i\partial\bar{\partial}h_1 = \omega$. By taking a sufficiently large b and an appropriate c, we may assume that h_1 is strictly increasing with respect to $|z_1|$ and satisfies $h_1(z_1, z_2) \leq 0$ for $|z_1| \leq 1-2\varepsilon$ and $h_1(z_1, z_2) \geq 1$ for $|z_1| \geq 1-\varepsilon$. Then we define $h_2 \colon H_a \to \mathbb{R}$ by

$$h_2(z_1, z_2) = \begin{cases} \max\{h_1(z_1, z_2), 0\} & (1 - 2\varepsilon \leq |z_1| \leq 1), \\ 0 & (0 \leq |z_1| \leq 1 - 2\varepsilon). \end{cases}$$

and $h_3 \colon H_a \to \mathbb{R}$ by

$$h_3(z_1, z_2) = d(|z_1|^2 + |z_2|^2),$$

where d is a sufficiently small positive number such that $h_3 < 1$ holds for $|z_1| \geq 1 - \varepsilon$. Then the function $h_4(z_1, z_2) = \max(h_2, h_3)$ is a continuous strictly plurisubharmonic function, since it coincides with $\max(h_1, h_3)$ where h_1 is defined, and with h_3 where is not. Moreover, it coincides with h_1 if $|z_1| \geq 1 - \varepsilon$ and with h_3 if $|z_1| \leq 1 - 2\varepsilon$. By Theorem 5.6, we can approximate h_4 from above by a smooth strictly plurisubharmonic function h_5. Then, the Kähler form $i\partial\bar{\partial}h_5$ is an extension of ω over H_a.   □

Then, we can make the concave holomorphic filling in Theorem 7.1 Kähler, since the complement of a small strongly pseudoconvex 4-ball in any compact Kähler surface is a Kähler concave holomorphic filling of (S^3, ξ_0), and an iteration of holomorphic handle attaching preserves the Kählerness if each handle is thin enough.

For any two positive closed connected contact 3-manifolds (M_1, ξ_1) and (M_2, ξ_2), Theorem 6.14 gives a sequence of contact (± 1)-surgeries from (M_1, ξ_1) to (M_2, ξ_2). Applying Lemma 7.2 to the sequence, we obtain a complex cobordism X from (M_1, ξ_1) to (M_2, ξ_2) that admits a Kähler metric. By Theorem 5.5, however, if (M_1, ξ_1) is Stein fillable and (M_2, ξ_2) is overtwisted, there is no symplectic cobordism from (M_1, ξ_1) to (M_2, ξ_2). In this case, the above X is Kähler as a complex manifold, but is not a symplectic cobordism. The cause of such a phenomenon is the incompatibility between the Kähler form extended in Lemma 7.2 and the boundary of the handle H_a. Indeed, the level sets of the Kähler potential h_4 are transverse to ∂H_a where $|z_1|$ is small. Therefore, the existence of an inward transverse Liouville vector field is not guaranteed and this prevents X from becoming a symplectic cobordism. To avoid such an obstacle, we introduce the following notion.

Definition 7.3 (Concave Kähler Filling) A strongly pseudoconcave surface (V, J) is called a *concave Kähler filling* of (M, ξ) if (V, J) is a concave holomorphic filling of (M, ξ), and there exists a Kähler form ω on V and a Kähler potential on a collar neighborhood of ∂V that is a defining function of V.

Since a concave Kähler filling is a concave symplectic filling, the above inconvenience has been resolved. Then, it is natural to regard this as the true definition of Kählerness for a strongly pseudoconcave surface and to ask the following question.

Problem 7.3 Which positive closed contact 3-manifold admits a concave Kähler filling?

We immediately see by Theorem 6.17 that if (M, ξ) is Stein fillable, then it admits a concave Kähler filling.

Example 7.3 Let W be a Stein filling of a Stein fillable contact 3-manifold (M, ξ). By Theorem 6.17, Int W is biholomorphic and symplectomorphic to a domain X in a projective surface S. Then, $S - X$ is a Kähler concave holomorphic filling of (M, ξ).

In consideration of this example, it would be interesting to ask whether every contact 3-manifold admitting a concave Kähler filling is Stein fillable. On the other hand, Takeo Ohsawa pointed out in some conference that the corresponding statement does not hold for CR 3-manifolds. In fact, the Rossi sphere is a counterexample.

Anyway, Problem 7.3 is a future issue and we are now investigating what kind of approach (contact geometric or complex geometric) is appropriate.

References

1. Akbulut, S., Ozbagci, B.: Lefschetz fibrations on compact Stein surfaces. Geom. Topol. **5**, 319–334 (2001)
2. Akhmedov, A., Etnyre, J., Mark, T., Smith, I.: A note on Stein fillings of contact manifolds. Math. Res. Lett. **15**(6), 1127–1132 (2008)
3. Alexander, J.: A lemma on systems of knotted curves. Proc. Nat. Acad. Sci. USA **9**, 93–95 (1923)
4. Arnol'd, V.I.: Mathematical Methods of Classical Mechanics. Graduate Texts in Mathematics, vol. 60. Springer, Berlin (1989)
5. Arnol'd, V.I., Givental', A.B.: Symplectic Geometry. Dynamical Systems IV. Encyclopaedia of Mathematical Sciences, vol. 4. Springer, Berlin (1990)
6. Barth, K., Geiges, H., Zehmisch, K.: The diffeomorphism type of symplectic fillings. J. Symplectic Geom. **17**, 929971 (2019)
7. Baykur, I., Van Horn-Morris, J.: Families of contact 3-manifolds with arbitrarily large Stein fillings. J. Differ. Geom. **101**(3), 423–465 (2015)
8. Bennequin, D.: Entrelacements et équations de Pfaff. Astérisque **107–108**, 87–161 (1983)
9. Bertin, J., Demailly, B., Illusie, L., Peters, C.: Introduction to Hodge Theory. AMS Texts and Monographs, vol. 8. American Mathematical Society, Providence (1996)
10. Bishop, E.: Mappings of partially analytic spaces. Am. J. Math. **83**, 209–242 (1961)
11. Bogomolov, F.: Classification of surfaces of class VII_0 with $b_2 = 0$. Math. USSR Izv. **10**, 255–269 (1976)
12. Bogomolov, F.: Surfaces of class VII_0 and affine geometry. Math. USSR Izv. **21**, 31–73 (1983)
13. Bogomolov, F., de Oliveira, B.: Stein small deformations of strictly pseudoconvex surfaces. Contemp. Math. Am. Math. Soc. **207**, 25–41 (1998)
14. Borman, M.S., Eliashberg, Y., Murphy, E.: Existence and classification of overtwisted contact structures in all dimensions. Acta Math. **215**, 281–361 (2015)
15. Boutet de Monvel, L.: Intégration des équations de Cauchy-Riemann induites formelles. Séminarie Équations aux dérivées partielles (Polytechnique), Exp. no. 9, pp. 1–13 (1974–1975)
16. Burns, D.M., Epstein, C.L.: Embeddability for three-dimensional CR-manifolds. J. Am. Math. Soc. **3–4**, 809–841 (1990)
17. Calabi, E., Eckmann, B.: A class of compact, complex manifolds which are not algebraic. Ann. Math. **58**, 494–500 (1953)

© The Editor(s) (if applicable) and The Author(s), under exclusive license
to Springer Nature Singapore Pte Ltd. 2025
N. Kasuya, *Non-Kähler Complex Surfaces and Strongly Pseudoconcave Surfaces*,
SpringerBriefs in Mathematics, https://doi.org/10.1007/978-981-96-3002-8

18. Cartan, H., Thullen, P.: Zur Theorie der Singularitäten der Funktionen mehrerer komplexen Veränderlichen. Regularitäts-und Konvergenzbereiche. Math. Ann. **106**, 617–647 (1932)
19. Castelnuovo, G., Enriques, F.: Sopra alcune questioni fondamentali nella teoria delle superficie algebriche. Ann. Mat. Pura Appl. Ser. III **6**, 165–225 (1901)
20. Chanillo, S., Chiu, H.L., Yang, P.: Embeddability for 3-dimensional Cauchy-Riemann manifolds and CR Yamabe invariants. Duke Math. J. **161**(15), 2909–2921 (2012)
21. Chanillo, S., Chiu, H.L., Yang, P. : Embedded three-dimensional CR manifolds and the non-negativity of Paneitz operators. In: Geometric Analysis, Mathematical Relativity, and Nonlinear Partial Differential Equations. Contemporary Mathematics, vol. 599, pp. 65–82. American Mathematical Society, Providence (2013)
22. Chern, S.S.: Characteristic classes of Hermitian manifolds. Ann. Math. **47**, 85–121 (1946)
23. Christian, A., Li, Y.: Some applications of menke's JSJ decomposition for symplectic fillings. Trans. Am. Math. Soc. **376**(7), 4569–4604 (2023)
24. Cieliebak, K., Eliashberg, Y.: From Stein to Weinstein and Back. Symplectic Geometry of Affine Complex Manifolds, vol. 59. AMS Colloqium Publications, Providence (2012)
25. Colin, V.: Chirurgies d'indice un et isotopies de sphères dans les variétés de contact tendues. C. R. Acad. Sci. Paris Sér. I Math. **324**, 659–663 (1997)
26. Conley, C., Zehnder, E.: The Birkhoff-Lewis fixed point theorem and a conjecture by V. I. Arnold. Invent. Math. **73**, 33–49 (1983)
27. Ding, F., Geiges, H.: A Legendrian surgery presentation of contact 3-manifolds. Math. Proc. Cambridge Philos. Soc. **136**, 583–598 (2004)
28. Ding, F., Geiges, H.: A unique decomposition theorem for tight contact 3-manifolds. Enseign. Math. **53**, 333–345 (2007)
29. Di Scala, A.J., Kasuya, N., Zuddas, D.: Non-Kähler complex structures on $\mathbb{R}^4$. Geom. Topol. **21**, 2461–2473 (2017)
30. Di Scala, A.J., Kasuya, N., Zuddas, D.: Non-Kähler complex structures on $\mathbb{R}^4$, II. J. Symplectic Geom. **16**, 631–644 (2018)
31. Dloussky, G.: Une construction élémentaire des surfaces d'Inoue-Hirzebruch. Math. Ann. **280**, 663–682 (1988)
32. Dolgachev, I.: On the link space of a Gorenstein quasihomogeneous surface singularity. Math. Ann. **265**, 529–540 (1983)
33. Ehlers, F., Neumann, W.D., Scherk, J.: Links of surface singularities and CR space forms. Comment. Math. Helv. **62**, 240–264 (1987)
34. Eliashberg, Y.: Classification of overtwisted contact structures on 3-manifolds. Invent. Math. **98**, 623–637 (1989)
35. Eliashberg, Y.: Topological characterization of Stein manifolds of dimension > 2. Int. J. Math. **1**, 29–46 (1990)
36. Eliashberg, Y.: Filling by holomorphic discs and its applications. In: Geometry of Low-Dimensional Manifolds, 2 (Durham, 1989), 45–67. London Mathematical Society Lecture Notes, vol. 151. Cambridge University Press, Cambridge (1990)
37. Eliashberg, Y.: Contact 3-manifolds twenty years since J. Martinet's work. Ann. Inst. Fourier **42**, 165–192 (1992)
38. Eliashberg, Y., Gromov, M.: Convex Symplectic Manifolds. Several Complex Variables and Complex Geometry, Part 2, pp. 135–162. American Mathematical Society, Providence (1989)
39. Eliashberg, Y., Gromov, M.: Embeddings of Stein manifolds of dimension n into the affine space of dimension $3n/2 + 1$. Ann. Math. **136**(1), 123–135 (1992)
40. Enoki, I.: Surfaces of class VII_0 with curves. Tohoku Math. J. **33**, 453–492 (1981)
41. Epstein C.L., Henkin, G.M.: Stability of embeddings for pseudoconcave surfaces and their boundaries. Acta Math. **185**(2), 161–237 (2000)
42. Etnyre, J.: Lectures on open book decompositions and contact structures. In: Floer Homology, Gauge Theory, and Low-Dimensional Topology. Clay Mathematics Institute, vol. 5, pp. 103–141. AMS, Providence (2006)
43. Etnyre, J., Honda, K.: On the nonexistence of tight contact structures. Ann. Math. **153**, 749–766 (2001)

44. Etnyre, J., Honda, K.: On symplectic cobordisms. Math. Ann. **323**, 31–39 (2002)
45. Etnyre, J., Roy, A.: Symplectic fillings and cobordisms of lens spaces. Trans. Am. Math. Soc. **374**(12), 8813–8867 (2021)
46. Floer, A.: Proof of Arnold conjecture for surfaces and generalizations to certain Kähler manifolds. Duke Math. J. **53**, 1–32 (1986)
47. Floer, A.: Morse theory for Lagrangian intersections. J. Differ. Geom. **28**, 513–547 (1988)
48. Floer, A.: Symplectic fixed points and holomorphic spheres. Commun. Math. Phys. **120**, 575–611 (1989)
49. Forstnerič, F.: Stein Manifolds and Holomorphic Mappings: The Homotopy Principle in Complex Analysis. A Series of Modern Surveys in Mathematics, vol. 56. Springer, Berlin (2017)
50. Forstnerič, F., Kozak, J.: Strongly pseudoconvex handlebodies. J. Korean Math. Soc. **40**(4), 727–745 (2003)
51. Friedman, R., Morgan, J.W.: Smooth Four-Manifolds and Complex Surfaces. Ergebnisse der Mathematik, vol. 27. Springer, Berlin (1994)
52. Fukaya, K., Ono, K.: Arnold conjecture and Gromov-Witten invariant. Topology **38**(5), 933–1048 (1999)
53. Geiges, H.: An Introduction to Contact Topology. Cambridge Studies in Advanced Mathematics, vol. 109. Cambridge University Press, Cambridge (2008)
54. Ghiggini, P., Lisca, P., Stipsicz, A.: Tight contact structures on some small Seifert fibered 3-manifolds. Am. J. Math. **129**(5), 1403–1447 (2007)
55. Giroux, E.: Convexité en topologie de contact. Comment. Math. Helv. **66**, 637–677 (1991)
56. Giroux, E.: Une infinité de structures de contact tendues sur une infinité de variétés. Invent. Math. **135**, 789–802 (1999)
57. Giroux, E.: Structures de contact en dimension trois et bifurcations des feuilletages de surfaces. Invent. Math. **141**, 615–689 (2000)
58. Giroux, E.: Contact geometry: from dimension three to higher dimensions. In: Proceedings of the International Congress of Mathematicians, Beijing, pp. 405–414 (2002)
59. Gompf, R.E.: Spin^c-structures and homotopy equivalences. Geom. Topol. **1**, 41–50 (1997)
60. Gompf, R.: Handle construction of Stein surfaces. Ann. Math. **148**, 619–693 (1998)
61. Gompf, R.E., Stipcitz, A.I.: 4-Manifolds and Kirby Calculus. Graduate Studies in Mathematics, vol. 20. American Mathematical Society, Providence (1999)
62. Grauert, H.: On Levi's problem and the imbedding of real-analytic manifolds. Ann. Math. **68**, 460–472 (1958)
63. Grauert, H.: Theory of q-convexity and q-concavity. In: Several Complex Variables, VII. Encyclopaedia of Mathematical Sciences, vol. 84, pp. 259–284. Springer, Berlin (1994)
64. Grauert, H., Remmert, R. : Theory of Stein Spaces. Classics Mathematics. Springer, Berlin (2004)
65. Griffith, P., Harris, J.: Principles of Algebraic Geomtery. Wiley, New York (1978)
66. Gromov, M.: Pseudo holomorphic curves in symplectic manifolds. Invent. Math. **82**, 307–347 (1985)
67. Gunning, R.C., Rossi, H.: Analytic functions of several complex variables. AMS Chelsea Publishing, Providence (2009)
68. Hatcher, A.: Vector Budles and K-theory (2017, preprint)
69. Hironaka, H.: Resolution of singularities of an algebraic variety over a field of characteristic zero I, II. Ann. Math. **79**, 109–203 (1964); (2) **79**, 205–326 (1964)
70. Hirzebruch, F.: Über eine Klasse von einfachzusammenhängenden komplexen Mannigfaltigkeiten. Math. Ann. **124**, 77–86 (1951)
71. Hirzebruch, F.: Some problems on differentiable and complex manifolds. Ann. Math. **60**, 213–236 (1954)
72. Hirzebruch, F.: Hilbert modular surfaces. Enseign. Math. **71**, 183–281 (1973)
73. Hofer, H.: Lusternik-Schnirelman-theory for Lagrangian intersections. Ann. Inst. H. Poincaré Anal. Non Linéaire **5**, 465–499 (1988)
74. Honda, K.: On the classification of tight contact structures I. Geom. Topol. **4**, 309–368 (2000)

75. Honda, K.: On the classification of tight contact structures II. J. Differ. Geom. **55**, 83–143 (2000)
76. Husemöller, D.: Fiber Bundles. Graduate Texts in Mathematics, vol. 20. Springer, New York (1994)
77. Inoue, M.: On surfaces of class VII_0. Invent. Math. **24**, 269–310 (1974)
78. Inoue, M.: New surfaces with no meromorphic functions. Proc. Int. Cong. Math. Vancouver **1**, 423–426 (1974)
79. Inoue, M.: An example of analytic surface (in Japanese). Sugaku **27**, 358–364 (1975)
80. Inoue, M.: New surfaces with no meromorphic functions, II. In: Complex Analysis and Algebraic Geometry, pp. 91–106. Iwanami Shoten/Cambridge University Press, Tokyo/Cambridge (1977)
81. Kanda, Y.: The classification of tight contact structures on the 3-torus. Commun. Anal. Geom. **5**, 413–438 (1997)
82. Karras, U.: Deformations of cusp singularities. Proc. Sympos. Pure Math. **30**, 37–40 (1977)
83. Kas, A.: Weierstrass normal forms and invariants of elliptic surfaces. Trans. Am. Math. Soc. **225**, 259–266 (1977)
84. Kas, A.: On the deformation types of regular elliptic surfaces. In: Complex Analysis and Algebraic Geometry, pp. 107–112. Iwanami Shoten/Cambridge University Press, Tokyo/Cambridge (1977)
85. Kasuya, N.: The canonical contact structure on the link of a cusp singularity. Tokyo J. Math. **37**(1), 1–20 (2014)
86. Kasuya, N.: On the links of simple singularities, simple elliptic singularities and cusp singularities. Demons. Math. **48**(2), 289–312 (2015)
87. Kasuya, N., Zuddas, D.: A concave holomorphic filling of an overtwisted contact 3-sphere. Algebraic Geom. Topol. **23**(5), 2141–2156 (2023)
88. Kasuya, N., Zuddas, D.: On the strongly pseudoconcave boundary of a compact complex surface. Proc. Am. Math. Soc. **152**(2), 709–723 (2024)
89. Kasuya, N., Kodama, H., Mitsumatsu, Y., Mori, A.: Lefschetz fibrations on the Milnor fibers of cusp and simple elliptic singularities (2021). arXiv:2111.00749
90. Kato, M.: Compact complex manifolds containing "global" spherical shells, I. In: Proceedings of the International Symposium on Algebraic Geometry, Kyoto, pp. 45–84 (1977)
91. Kato, M.: Compact complex surfaces containing global strongly pseudoconvex hypersurfaces. Tohoku Math. J. **31**, 537–547 (1979)
92. Kiremidjian, G.K.: Extendible pseudocomplex structures. J. Approx. Theory **19**(4), 281–303 (1977)
93. Kiremidjian, G.K.: A direct extension method for CR structures. Math. Ann. **242**(1), 1–19 (1979)
94. Klein, F.: Lectures on the Icosahedron and the Solution of the Fifth Degree. Teubner, Leipzig (1884); Dover, Mineola (1956)
95. Kodaira, K.: The theorem of Riemann-Roch on compact analytic surfaces. Am. J. Math. **73**, 813–875 (1951)
96. Kodaira, K.: On compact complex analytic surfaces: I. Ann. Math. **71**, 111–152 (1960)
97. Kodaira, K.: On compact analytic surfaces: II. Ann. Math. **77**, 563–626 (1963)
98. Kodaira, K.: On compact analytic surfaces: III. Ann. Math. **78**, 1–40 (1963)
99. Kodaira, K.: On the structures of compact complex analytic surfaces: I. Am. J. Math. **86**, 751–798 (1964)
100. Kodaira, K.: On the structures of compact complex analytic surfaces: II. Am. J. Math. **88**, 682–721 (1966)
101. Kodaira, K.: On the structures of compact complex analytic surfaces: III. Am. J. Math. **90**, 55–83 (1968)
102. Kodaira, K.: On the structures of compact complex analytic surfaces: IV. Am. J. Math. **90**, 1048–1066 (1968)
103. Kohn, J.J.: Harmonic integrals on strongly pseudo-convex manifolds. I. Ann. Math. **78**, 112–148 (1963)

104. Kohn, J.J.: The range of the tangential Cauchy-Riemann operator. Duke Math. J. **53**(2), 525–545 (1986)
105. Kohn, J.J., Rossi, H.: On the extension of holomorphic functions from the boundary of a complex manifold. Ann. Math. **81**, 451–472 (1965)
106. Lattès, M.S.: Sur les formes réduites des transformations ponctuelles à deux variables. Comptes Rendus **152**, 1566–1569 (1911)
107. Laufer, H.B.: On minimally elliptic singularities. Am. J. Math. **99**, 1257–1295 (1977)
108. Lazarev, O.: Contact manifolds with flexible fillings. Geom. Funct. Anal. **30**(1), 188–254 (2020)
109. Lempert, L.: On three-dimensional Cauchy-Riemann manifolds. J. Am. Math. Soc. **5**, 923–969 (1992)
110. Lempert, L.: Algebraic approximations in analytic geometry. Invent. Math. **121**, 335–354 (1995)
111. Li, J., Yau, S.T., Zheng, F.: On projectively flat Hermitian manifolds. Commun. Anal. Geom. **2**, 103–109 (1994)
112. Liouville, J.: Note sur l'intégration des équations différentielles de la Dynamique, présentée au Bureau des Longitudes le 29 juin 1853. J. Math. Pures Appl. Sér. 1 Tome **20**, 137–138 (1855)
113. Lisca, P.: On symplectic fillings of lens spaces. Trans. Am. Math. Soc. **360**, 765–799 (2008)
114. Lisca, P., Matić, G.: Tight contact structures and Seiberg-Witten invariants. Invent. Math. **129**, 509–525 (1997)
115. Loi, A., Piergallini, R.: Compact Stein surfaces with boundary as branched covers of B^4. Invent. Math. **143**, 325–348 (2001)
116. Looijenga, E.: Isolated Singular Points on Complete Intersections. London Mathematical Society Lecture Note Series, vol. 77. Cambridge University Press, Cambridge (1984)
117. Martinet, J.: Formes de contact sur les variétés de dimension 3. In: Proceedings of Liverpool Singularity Sympos, II. Lecture Notes in Mathematics, vol. 209, pp. 142–163. Springer, Berlin (1971)
118. Matsumoto, Y.: On 4-manifolds fibered by tori. Proc. Japan Acad. Ser. A Math. Sci. **58**(7), 298–301 (1982)
119. Matsumoto, Y.: Topology of torus fibrations (in Japanese). Sugaku **36**(4), 289–301 (1984)
120. Mastumoto, Y.: Diffeomorphism types of elliptic surfaces. Topology, **25**(4), 549–563 (1986)
121. McDuff, D.: Symplectic manifolds with contact type boundaries. Invent. Math. **103**(3), 651–671 (1991)
122. McDuff, D., Salamon, D.: J-Holomorphic Curves and Quantum Cohomology. AMS University Lecture Series, vol. 6. AMS, Providence (1994)
123. McDuff, D., Salamon, D.: Introduction to symplectic topology. Oxford Mathematical Monographs. Oxford University Press, Oxford (1995)
124. Milnor, J.W.: Morse Theory. Annals of Mathematics Studies, vol. 51. Princeton University Press, Princeton (1963)
125. Milnor, J.W.: Singular Points of Complex Hypersurfaces. Annals of Mathematics Studies, vol. 61. Princeton University Press, Princeton (1968)
126. Milnor, J.W.: On the 3-dimensional Brieskorn manifolds $M(p, q, r)$. Annals of Mathematics Studies, vol. 84, pp. 175–225 (1975)
127. Milnor, J.W., Stasheff, J.D.: Characteristic Classes. Annals of Mathematics Studies, vol. 76. Princeton University Press, Princeton (1974)
128. Miranda, R.: The moduli of Weierstrass fibrations over P^1. Math. Ann. **225**, 379–394 (1981)
129. Mitsumatsu, Y.: Topology of 3-dimensional contact structures (in Japanese). MSJ Memoir. **1**, 131 (2001)
130. Miyaoka, Y.: Kähler metrics on elliptic surfaces. Proc. Japan Acad. Ser. A Math. Sci. **50**, 533–536 (1974)
131. Moishezon, B.: Complex Surfaces and Connected Sums of Complex Projective Planes. Lecture Notes in Mathematics, vol. 603. Springer, Berlin (1977)

132. Mori, A.: Reeb foliations on S^5 and contact 5-manifolds violating the Thurston-Bennequin inequality (2009, preprint). arXiv:0906.3237v3 [math.GT]

133. Mori, A.: The Reeb foliation arises as a family of Legendrian submanifolds at the end of a deformation of the standard S^3 in S^5. Comp. Rendus Math. **350**(1–2), 67–70 (2012)

134. Nakamura, I.: Inoue-Hirzebruch surfaces and a duality of hyperbolic unimodal singularities. I. Math. Ann. **252**(3), 221–235 (1980)

135. Nakamura, I.: On surfaces of class VII_0 with curves. Invent. Math. **78**, 393–443 (1984)

136. Nakamura, I.: Towards classification of nonkählerian complex surfaces. Sugaku Exp. Math. **2**, 209–229 (1989)

137. Narasimhan, R.: Imbedding of holomorphically complete complex spaces. Am. J. Math. **82**, 917–934 (1960)

138. Narasimhan, R.: A note on Stein spaces and their normalisations. Ann. Scuola Norm. Sup. Pisa Cl. Sci. **16**, 327–333 (1962)

139. Neumann, W.D.: Geometry of quasihomogeneous surface singularities. Proc. Symp. Pure Math. **40**, 245–258 (1983)

140. Nirenberg, L.: Lectures on Linear Partial Differential Equations. Expository Lectures from the CBMS Regional Conference held at the Texas Technological University, Lubbock, May 22–26, 1972. Conference Board of the Mathematical Sciences Regional Conference Series in Mathematics, No. 17. American Mathematical Society, Providence (1973)

141. Nirenberg, L.: A certain problem of Hans Lewy (in Russian). Translated from the English by Ju. V. Egorov. Collection of articles dedicated to the memory of Ivan Georgievic Petrovskii (1901–1973), I. Uspehi Mat. Nauk **29**(2), 241–251 (1974)

142. Oba, T.: Higher-dimensional contact manifolds with infinitely many Stein fillings. Trans. Am. Math. Soc. **370**(7), 5033–5050 (2018)

143. Ohsawa, T.: Analysis of Several Complex Variables. Translations of Mathematical Monographs. Iwanami Series in Modern Mathematics, vol. 211. AMS, Providence (2002)

144. Ohsawa, T.: A survey on the embeddings of strongly pseudoconvex CR structures. In: Contact Structures, Singularities, Differential Equations and Related Topics (2022). https://drive. google.com/file/d/1fM_35K9gg60JZh0Vqqjy0i6v1YBvZvG_/view

145. Ohta, H., Ono, K.: Simple singularities and topology of symplectically filling 4-manifolds. Comment. Math. Helv. **74**, 575–590 (1999)

146. Ohta, H., Ono, K.: Symplectic fillings of the link of simple elliptic singularities. J. Reine Angew. Math. **565**, 183–205 (2003)

147. Ohta, H., Ono, K.: Simple singularities and symplectic fillings. J. Differ. Geom. **69**(1), 1–42 (2005)

148. Ono, K.: On the Arnol'd conjecture for weakly monotone symplectic manifolds. Invent. Math. **119**(3), 519–537 (1995)

149. Orlik, P., Wagreich, P.: Isolated singularities of algebraic surfaces with $\mathbb{C}^*$ action. Ann. Math. **93**, 205–228 (1971)

150. Ozbagci, B., Stipsicz, A.I.: Contact 3-manifolds with infinitely many Stein fillings. Proc. Am. Math. Soc. **132**(5), 1549–1558 (2004)

151. Ozbagci, B., Stipsicz, A.I.: Surgery on Contact 3-Manifolds and Stein Surfaces. Bolyai Society Mathematical Studies, vol. 13. Springer, Berlin (2004)

152. Peixoto, M.M.: Structual stability on two-dimensional manifolds. Topology **1**, 101–120 (1962)

153. Peixoto, M.M.: Structual stability on two-dimensional manifolds - a further remark. Topology **2**, 179–180 (1963)

154. Phillips, A.: Submersions of open manifolds. Topology **6**, 171–206 (1967)

155. Pinkham, H.: Normal surface singularities with $\mathbb{C}^*$-action. Math. Ann. **227**, 183–193 (1977)

156. Remmert, R.: Sur les espaces analytiques holomorphiquement séparables et holomorphiquement convexes. C. R. Acad. Sci. Paris **243**, 118–121 (1956)

157. Richberg, R.: Stetige streng pseudokonvexe Funktionen. Math. Ann. **175**, 251–286 (1968)

158. Rossi, H.: Attaching analytic spaces to an analytic space along a pseudoconcave boundary. In: Proceedings of Conference on Complex Analysis (Minneapolis, 1964), pp. 242–256. Springer, Berlin (1965)

159. Saito, K.: Quasihomogene isolierte Singularitäten von Hyperflächen. Invent. Math. **14**, 123–142 (1971)

160. Saito, K.: Einfach-elliptische Singularitäten. Invent. Math. **23**, 289–325 (1974)

161. Sakamoto, K., Fukuhara, S.: Classification of T^2-bundles over T^2. Tokyo J. Math. **6**(2), 311–327 (1983)

162. Seade, J.: On the Topology of Isolated Singularities in Analytic Spaces. Progress in Mathematics, vol. 241. Birkhäuser, Basel (2006)

163. Serre, J.P.: Un théorème de dualité. Comment. Math. Helv. **29**, 9–26 (1955)

164. Siegel, C.L.: Meromorphe Funktionen auf kompakten analytischen Mannigfaltigkeiten. Nachr. Akad. Wiss. Göttingen **4**, 71–77 (1955)

165. Siu, Y.T.: Every $K3$ surface is Kähler. Invent. Math. **73**, 139–150 (1983)

166. Steenrod, N.E.: Topology of Fiber Bundles. Princeton University Press, Princeton (1951)

167. Sternberg, S.: Local contractions and a theorem of Poincaré. Am. J. Math. **79**, 809–824 (1957)

168. Stipsicz, A.I.: Gauge theory and Stein fillings of certain 3-manifolds. In: Proceedings of the Gökova Geometry-Topology Conference, pp. 115–131 (2001)

169. Takeuchi, Y.: Nonnegativity of the CR Paneitz operator for embeddable CR manifolds. Duke Math. J. **169**(18), 3417–3438 (2020)

170. Teleman, A. : Projectively flat surfaces and Bogomolov's theorem on class VII_0-surfaces. Int. J. Math. **5**(2), 253–264 (1994)

171. Teleman, A.: Donaldson theory on non-Kählerian surfaces and class VII surfaces with $b_2 = 1$. Invent. Math. **162**, 493–521 (2005)

172. Teleman, A.: Instantons and holomorphic curves on class VII surfaces. Ann. Math. **172**, 1749–1804 (2010)

173. Teleman, A.: Donaldson theory in non-Kählerian geometry. In: Modern Geometry: A Celebration of the Work of Simon Donaldson, Proceedings of Symposia in Pure Mathematics. AMS, Providence, pp. 363–392 (2018)

174. Thurston, W., Winkelnkemper, H.: On the existence of contact forms. Proc. Am. Math. Soc. **52**, 345–347 (1975)

175. Voisin, C.: Hodge Theory and Complex Algebraic Geometry, I, II. Cambridge Studies in Advanced Mathematics, vols. 76–77. Cambridge University Press, Cambridge (2007)

176. Wagreich, P.: The structure of quasihomogeneous singularities. Proc. Sympos. Pure Math. **40**, 593–611 (1983)

177. Wendl, C.: Strongly fillable contact manifolds and J-holomorphic foliations. Duke Math. J **151**(3), 337–384 (2010)

178. Whitney, H.: Analytic extensions of differentiable functions defined in closed sets. Trans. Am. Math. Soc. **36**(1), 63–89 (1934)